SCIENCE
A CLOSER LOOK

BUILDING SKILLS
Assessment

Mc
Graw
Hill
Education

Contents

Contents

Macmillan/McGraw-Hill *Assessment* in science, Grade 3, is a comprehensive program designed to familiarize students with standardized testing in science and to review the concepts covered in Macmillan/McGraw-Hill *Science: A Closer Look*. The practice tests and performance assessment activities in this book can also serve as tools in a complete program of assessment to help gauge mastery of the science content students have learned.

About This Book

The questions in this book will accustom students to standardized testing in science, including multiple-choice and open-response style questions about Life Science, Earth Science, and Physical Science, in a grade-appropriate manner. General scientific methods are stressed along with critical thinking.

The main components of this book coincide with the respective chapters and lessons in Macmillan/McGraw-Hill *Science: A Closer Look* and include:

- **Chapter Tests A and B:** Each summative practice test covers science content from the corresponding chapters and tests students' knowledge of important vocabulary and concepts they have learned. Key concepts are tested in several ways to ensure that students comprehend core content. Skills such as making inferences, drawing conclusions, and scientific thinking are emphasized in the practice tests. Both practice tests cover the same content, but test the material in different ways, providing the teacher with several options of using the tests as pretests and posttests, chapter tests, homework assignments, or as extra practice.

- **Lesson Tests:** These pages provide test practice and focus on specific concepts covered in each lesson of the corresponding chapter.

- **Performance Assessment Activity:** Each activity covers a main concept from the corresponding chapter and provides students with a hands-on exercise that further reinforces the content they have learned. A rubric precedes each activity and provides guidelines for grading students' performance. Performance assessment activities require adult supervision.

How to Administer the Practice Tests

- Remove the practice test pages from the book and photocopy them for students. Answers for all questions are marked in non-reproducible blue ink.

- Separate students' desks so that students can work independently.

- Tell students that they are taking a practice test and ask them to remove everything from their desks except for several pencils. They may not speak to classmates until the test is over.

- Keep the classroom atmosphere as much like the administration of a standardized test as possible. Minimize distractions and discourage talking.

The scientific knowledge assessed in this book and in Macmillan/McGraw-Hill *Science: A Closer Look* will help students build a strong foundation in science and lay the groundwork for future learning.

A Look at Living Things

Write the word that best completes each sentence in the spaces below. Words may be used only once.

cells	invertebrate	reptiles	vertebrates
environment	organisms	shelter	
exoskeleton	photosynthesis	structures	

1. All living things are _____organisms_____ .

2. Everything that surrounds a living thing is called a(n) _____environment_____ .

3. An animal that has no backbone is called a(n) _____invertebrate_____ .

4. Plants and animals have _____structures_____ that help them get what they need to survive.

5. Animals that have backbones are called _____vertebrates_____ .

6. The process by which a plant makes food is called _____photosynthesis_____ .

7. A place in which an animal can stay safe is called a(n) _____shelter_____ .

8. A thin, hard covering that surrounds certain animals' bodies is called a(n) _____exoskeleton_____ .

9. The building blocks of life are called _____cells_____ .

10. Vertebrates that have scaly skin and breathe through lungs are called _____reptiles_____ .

Name _____ Date _____

Circle the letter of the best answer for each question.

11. What type of animal has fur and breathes using lungs?

   A amphibian

   B bird

   C mammal

   D fish

12. What is one way plants and animals are different from each other?

   A Plants never stop growing.

   B Plants need water.

   C Animals make their own food.

   D Animals need nutrients.

13. Which is how living things make more of their own kind?

   A photosynthesis

   B reproduction

   C response

   D survival

14. What substances help living things grow and stay healthy?

   A nutrients

   B structures

   C shelters

   D environments

15. Which animal spends part of its life in water and part on land?

   A amphibian

   B bird

   C mammal

   D fish

© Macmillan/McGraw-Hill

Answer the following questions.

16. **Classify** Study the list of animals. Identify each type of vertebrate based on its characteristics.

amphibians	birds	fish	reptiles

strong muscles; lightweight bones: _____ birds

breathe through gills; spend entire lives in the water: _____ fish

scaly skin; can live on land or water: _____ reptiles

start out with gills; grow lungs to live on land: _____ amphibians

17. **Infer** Why do plants make their own food?

Plants make their own food because they cannot move

around to find food like animals do.

18. **Make a Model** Draw and label the three common structures found on a plant.

Students' drawing should include the following: roots—either

one large taproot (as in a carrot or other root vegetable)

or a web of thinner roots; stem—may include a traditional

plant stem or a large stem such as a tree trunk; leaves—may

include leaves or tree needles

19. **Define** How are lungs and gills similar?

Lungs are animal structures that take in oxygen from the air.

Gills are animal structures that take in oxygen from the water.

Both take in oxygen.

Answer the following questions.

20. **Communicate** Fill in the chart below by naming three characteristics that set mammals apart from other vertebrates. Then tell how each characteristic is useful to the mammal.

Characteristic	How It Is Useful
live births; mothers nursing their young; mothers take care of young until they can take care of themselves; all mammals have fur or hair	reduces risk of death during incubation period; constant supply of food, no need to hunt for food; provides more protection against predators; protection from the elements

21. **Critical Thinking** What accounts for the difference in plants around the world?

Plants grow in different environments. Some places are

dry and hot, other places are snowy and cold. Plants have

structures that work best in their environments.

22. **Thinking Like a Scientist** Describe what characteristics would help you classify a turtle and a rabbit.

Students' answers will vary but may include the following:

turtles have smooth skin, lay eggs that hatch into young; live

in and/or near water, so they are reptiles; rabbits have fur,

give birth to live young, provide milk as food for their young,

so they are mammals.

Circle the letter of the best answer for each question.

1. Which of these items belongs in the empty circle?

A safety
B sunlight
C space
D sleep

2. What is the special tool that makes small things look larger?

A projector
B microscope
C periscope
D microphone

3. What is another name for "changing with age"?

A growing
B responding
C energizing
D reproducing

4. What happens when a plant responds to sunlight?

A it bends toward the light
B it retains water
C it covers itself with flowers
D it moves underground

Critical Thinking What kinds of plants and animals would most likely survive in a desert environment?

Plants and animals that do well in hot climates and can survive on

little water could live in the desert.

Circle the letter of the best answer for each question.

1. What is the name for the substances that help living things grow and stay healthy?

   A roots

   B nutrients

   C structures

   D energy

2. Which substance gives leaves their green color?

   A oxygen

   B carbon dioxide

   C chlorophyll

   D photosynthesis

3. Needles on an evergreen tree are examples of which plant structure?

   A leaf

   B stem

   C root

   D flower

4. Which of the following is not a characteristic of a stem?

   A holds up a plant

   B produces nutrients

   C carries water and food

   D holds leaves in place

Critical Thinking Why is it important for people to preserve the world's rain forests?

The rain forests are filled with trees and plants. Trees and plants

give off oxygen that people need. By cutting down rain forests,

people are decreasing the world's oxygen supply.

_____

© Macmillan/McGraw-Hill

Name _____ Date _____ Date _____ Name

Name _____ Date _____

Name _____ **Date** _____

Lesson 3 Test

Circle the letter of the best answer for each question.

1. Which of the following is an example of a structure that helps keep an animal safe?

   (A) quills

   B gills

   C nests

   D burrows

2. What are lungs?

   (A) animal structures that take in oxygen from the air

   B plant structures that carry water through a plant

   C animal structures that take in oxygen from the water

   D plant structures that give plants their color

3. What is one thing all animals have in common?

   A All animals have fur.

   B All animals breathe through lungs.

   C All animals can walk.

   (D) All animals eat other organisms for food.

4. How do worms take in oxygen?

   A through the air

   (B) through their skin

   C through the water

   D through sunlight

Critical Thinking Compare a bear with a giraffe. How do their structures help them survive?

A bear has sharp teeth and claws to help it eat. A bear also has

thick fur and a lot of body fat to help it survive cold winters. A

giraffe has long legs to outrun predators and a long neck to help it

reach leaves that are high in trees.

© Macmillan/McGraw-Hill

Circle the letter of the best answer for each question.

1. Most of the animals on Earth are

   A vertebrates.

   (**B**) invertebrates.

   C arthropods.

   D mollusks.

2. Which of the following is an example of an invertebrate?

   A lizard

   B bird

   (**C**) sea star

   D goldfish

3. What does it mean to classify an animal?

   A to capture it for scientific research

   (**B**) to put it into a group with other animals that are similar

   C to release it into the wild

   D to study it in its natural environment

4. Clams, snails, and octopuses are part of which group of invertebrates?

   (**A**) mollusks

   B arthropods

   C sponges

   D urchins

Critical Thinking What characteristics do lions have that help scientists classify them?

A lion is a mammal. It is a meat-eating predator. It has fur and
walks on four legs. It lives in Africa.
_____
_____
_____

A Look at Living Things

Write the word that best completes each sentence in the spaces below. Words may be used only once.

cells	invertebrates	reptiles	vertebrates
environment	organisms	shelter	
exoskeleton	photosynthesis	structures	

1. Some animals' bodies are surrounded by a thin, hard covering called a(n) _____exoskeleton_____ .

2. All living things are made of building blocks called _____cells_____ .

3. Scaly-skinned vertebrates that breathe through lungs are called _____reptiles_____ .

4. All plants go through a food-making process called _____photosynthesis_____ .

5. Animals stay safe in a place called a(n) _____shelter_____ .

6. Animals and plants get what they need to survive through different _____structures_____ .

7. Animals that have backbones are _____vertebrates_____ .

8. All living things can be described as _____organisms_____ .

9. Animals without backbones are _____invertebrates_____ .

10. All living things are surrounded by a(n) _____environment_____ .

Name _____ Date _____

Circle the letter of the best answer for each question.

11. How do trees make food?

  A response

  B survival

  C reproduction

  D photosynthesis

12. How do plants and animals differ?

  A Animals need nutrients.

  B Animals make their own food.

  C Animals cannot make their own food.

  D Plants need water.

13. Which animal has gills and cannot survive out of water?

  A mammal

  B amphibian

  C fish

  D bird

14. What do nutrients do?

  A help living things grow and stay healthy

  B give shelter to living things

  C make food from sunlight

  D take in water from the ground

15. Which of these animals look like fish when they hatch?

  A birds

  B reptiles

  C mammals

  D amphibians

Answer the following questions.

16. **Define** Birds take in oxygen from the air. Fish take in oxygen from the water. Label each animal with the correct structure for taking in oxygen.

lung _____   gill _____

17. **Infer** Why would a plant not grow in a dark room?

Plants' leaves trap energy from the sunlight to help them grow.

18. **Make a Model** Trees have common structures found on all plants. Draw and label a tree's roots, stem, and leaves.

Students' drawings should include: a web of thin roots; a tree

trunk and leaves (or tree needles)

19. **Classify** For each animal listed in the chart, name one characteristic that is specific only to that animal.

Animal	Characteristic
frog	start out with gills; grow lungs to live on land
fish	breathe through gills; spend entire lives in the water

Answer the following questions.

20. **Communicate** Name three characteristics of mammals. Describe how each characteristic is useful to the mammal.

Characteristics of Mammals

Mammals have live births, which reduces the risk of death during incubation period.

Mothers take care of their babies until they can take care of themselves, which means babies are safer from predators.

Mammals nurse their young, which provides a constant supply of food so babies do not have to hunt for food.

21. **Critical Thinking** Anne wants to send a friend in Alaska a plant. What factors should she consider when choosing a plant?

Temperature and amount of sunlight are two factors to

consider. She should choose a plant that can thrive in the cold

and does not need a lot of sunlight.

22. **Thinking Like a Scientist** Tell how you would classify an alligator.

Alligators have scaly skin, lay eggs that hatch into young, live in

and/or near water, so they are reptiles.

Living Together

Objective: Student will demonstrate an understanding of plant and animal needs and structures by constructing a two-dimensional simulated environment.

Materials

- poster board
- markers or colored pencils
- old magazines (for cutting out pictures)
- glue
- scissors

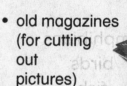

Scoring Rubric

4 points Student constructs a detailed and accurate representation of all of the words listed. Student includes various plants and animals that are well-suited to the scene. Student clearly labels all plants, animals, and structures. Student correctly explains his or her answers to the questions in Analyze the Results.

3 points Student shows a reasonably accurate representation of most of the words listed. Student includes mostly plants and animals that are well-suited to the scene. Student labels most of the plants, animals, and structures listed. Student gives partial explanation of his or her answers to the questions in Analyze the Results.

2 points Student shows a partially accurate representation of most of the words listed. Student includes few plants and animals that are well-suited to the scene, or may include some inaccurately placed species. Student labels a few of the plants, animals, and structures accurately. Student gives mostly incomplete explanation of his or her answers to the questions in Analyze the Results.

1 point Student shows a mostly inaccurate representation of the words listed. Student does not include plants and animals that are well-suited to the scene. Student does not label plants, animals, and structures, or labels them incorrectly. No explanation of his or her answers to the questions in Analyze the Results is given.

Living Together

Communicate

Use the library, media center, or the Internet to find a scene from nature. Draw or cut out pictures from old magazines of plants and animals that could live in that environment. Then identify and label the plants and animals using the words listed below.

amphibians	gills	mammal	shelter
birds	invertebrate	reptile	stem
fish	leaves	root	vertebrate

Analyze the Results

1. What does your environment need in order for the plants and animals to survive?

   Answers will vary but may include the following: Students

   report that the environment needs a source of water and

   sunlight for the plants and a source of food, water, sunlight,

   and shelter for the animals.

2. Choose one species of animal that you have included in your environment. Research that animal and describe what it needs to survive.

   Answers will vary depending on the animal selected. Answers

   should include discussion of food, shelter, climate, and

   habitat.

© Macmillan/McGraw-Hill

Living Things Grow and Change

Write the word or words that best complete each sentence in the spaces below. Words may be used only once.

cones	larva	pollination	traits
heredity	life cycle	pupa	
inherited traits	metamorphosis	seed	

1. Passing traits from parents to their young is called

   _____heredity_____ .

2. A process through which an organism's body

   changes form is called _____metamorphosis_____ .

3. When an insect hatches, it is called a(n) _____larva_____ .

4. Traits that are passed down from parents to their

   children are called _____inherited traits_____ .

5. The stage in an insect's life cycle between larva

   and adult is called the _____pupa_____ .

6. Features of a living thing, are called _____traits_____ .

7. Conifers use _____cones_____ to reproduce.

8. A(n) _____life cycle_____ is the stages in an

   organism's life.

9. The process of moving pollen from the male part

   of a flower to the female part of a flower is _____pollination_____ .

10. A new plant grows from a(n) _____seed_____ .

Circle the letter of the best answer for each question.

11. When a seed is planted in the soil and the conditions are right, it begins to grow in a process called

  A pollination.

  B germination.

  C fertilization.

  D decomposition.

12. What is the first stage in the life cycle of a ladybug?

  A larva

  B adult

  C egg

  D pupa

13. Riding a bicycle is an example of what kind of trait?

  A learned trait

  B inherited trait

  C heredity trait

  D environmental trait

14. What three factors affect an organism's traits?

  A heredity, growth, learning

  B heredity, environment, learning

  C environment, temperature, heredity

  D environment, sunlight, location

15. What is the larval stage of a frog called?

  A pupa

  B caterpillar

  C egg

  D tadpole

Answer the following questions.

16. Sequence Use the chart below to list the stages of a beetle's life cycle in the correct order.

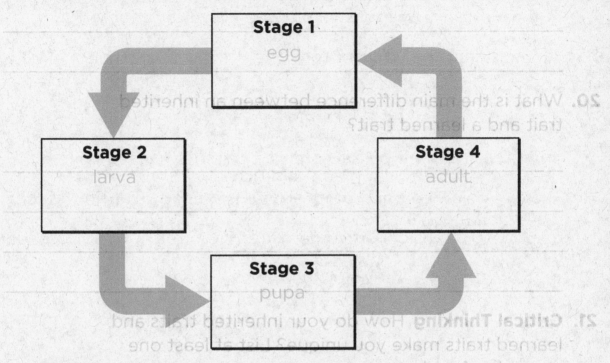

Stage 1
egg

Stage 2
larva

Stage 4
adult

Stage 3
pupa

17. Predict What changes does a caterpillar go through before becoming a butterfly?

The caterpillar will eat and grow. The caterpillar will then

enter the pupal stage in which it will form a cocoon. While

in the pupal stage, the caterpillar will transform and emerge

from the cocoon as a butterfly.

18. Infer How is the egg stage helpful to a turtle's life cycle?

The young turtle can grow safely inside the egg. The turtle will

get everything it needs from the egg and the shell will protect

it from many dangers.

Answer the following questions.

19. Describe two ways in which a plant can be pollinated.

The wind can blow pollen from one flower to another.

Animals can carry pollen from one flower to another.

20. What is the main difference between an inherited trait and a learned trait?

Inherited traits are passed on from parents to their children,

like eye color. Learned traits are skills gained over time, such

as playing a musical instrument. Learned traits are not passed

on from one generation to the next.

21. Critical Thinking How do your inherited traits and learned traits make you unique? List at least one example of each.

Sample answer: I am unique because I have red hair, which is

an inherited trait. I am also a good soccer player, which is a

learned trait.

22. Thinking Like a Scientist How could the insect population of an area have an effect on the growth of plants and flowers?

The insect population may help the plants and flowers grow

by pollination. However, some of the insects might eat the

plants, which could reduce the plant population.

© Macmillan/McGraw-Hill

Circle the letter of the best answer for each question.

1. What is another name for a young plant?

   (A) embryo

   B seed

   C leaf

   D bud

2. What happens to a seed when it germinates?

   A it dies

   (B) it begins to grow

   C it blooms

   D it takes in water

egg

   ↓

3. Which two parts of a flower help it make seeds?

   A roots and stems

   B seeds and seedlings

   (C) pollen and eggs

   D fruit and nectar

4. What has to happen before a seed can germinate?

   A it has to have sunlight

   B it has to be carried by wind

   C it has to have fruit

   (D) it has to be planted in soil

Critical Thinking How can a flower's sweet smell help it to become pollinated?

A flower's sweet smell will attract insects. The insects will fly from

one flower to another, carrying pollen. The insect will transfer the

pollen from one flower to another, which pollinates the flowers.

_____

Circle the letter of the best answer for each question.

1. What is another name for a young frog?

   A a pupa

   B a tadpole

   C a toad

   D a chrysalis

2. The hard case in which the pupa lives is

   A the nest.

   B the shell.

   C the cocoon.

   D the egg.

3. How are birds, fish, and reptiles alike?

   A They all have legs.

   B They all breathe through lungs.

   C They are all born looking like their parents.

   D They all begin their lives as eggs.

4. This chart shows the stages in a ladybug's life cycle. Which belongs in the empty box?

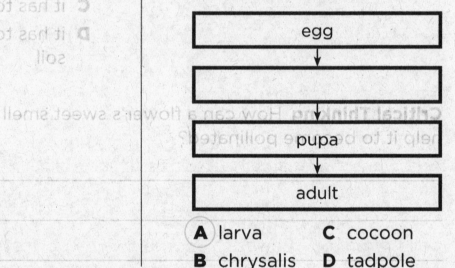

   | egg |
   | |
   | pupa |
   | adult |

   A larva   **C** cocoon

   B chrysalis   **D** tadpole

Critical Thinking Why do mammals care for their young until they can get food on their own, and reptiles and fish do not?

Reptiles and fish are born knowing how to get food. Mammals'

babies need to be taught how to catch or find their own food.

Circle the letter of the best answer for each question.

1. What is another name for an organism's young?

   A egg

   B seedling

   C offspring

   D pupa

2. What is an example of a learned trait?

   A a flower blooming

   B a bird eating worms

   C a bear hibernating

   D a person using a hammer

3. Which of the following is not an inherited trait?

   A a flower's shape

   B a person's soccer skills

   C a person's eye color

   D a dog's fur color

4. Look at the chart below. Which of the following traits belongs in the empty box?

Inherited	Learned
eye color	riding a bike
skin tone	speaking a language
	playing the flute

   A favorite color is blue

   B curly brown hair

   C likes to sing

   D loves spaghetti

Critical Thinking Why is it that children born of the same parents do not look exactly alike?

Children inherit certain traits from their parents. However, each

child may pick up different traits from each parent. One child

may have the father's traits for skin tone and hair color and the

mother's traits for eye color and height.

Living Things Grow and Change

Write the word or words that best complete each sentence
in the spaces below. Words may be used only once.

cones	larva	pollination	traits
heredity	life cycle	pupa	
inherited traits	metamorphosis	seed	

1. The _____ is the stage of an insect's life
   cycle in which the insect hatches.

2. Moving pollen from the male part of a flower to
   the female part is called _____ .

3. A larva transforms into a _____ .

4. A plant embryo grows inside a _____ .

5. Another name for the passing of traits from
   parents to their offspring is _____ .

6. An organism's _____ is the completion
   of all the stages in its life.

7. Traits that pass from parents to offspring are
   called _____ .

8. An organism's body changes during _____ .

9. Plant structures that make seeds are _____ .

10. The shape of a plant's flowers and leaves are
    features called _____ .

Circle the letter of the best answer for each question.

11. What is the second stage in a beetle's life cycle?

   A adult

   B egg

   C larva

   D pupa

12. Riding a bicycle and speaking a language are

   A heredity traits.

   B environmental traits.

   C learned traits.

   D inherited traits.

13. Animals such as bees carry pollen from one plant to another in a process called

   A germination.

   B decomposition.

   C pollination.

   D communication.

14. What is the name for the larval stage of a butterfly?

   A caterpillar

   B egg

   C tadpole

   D pupa

15. Which of the following is an inherited trait?

   A hair color

   B tanned skin

   C red dress

   D small scar

Answer the following questions.

16. Infer How is the egg stage of a bird similar to the pupal stage of a butterfly?

The baby bird gets everything it needs to survive from the

egg. The bird will hatch from its egg. Like the baby bird in the

egg, the caterpillar gets everything it needs to change into a

butterfly. The butterfly will emerge from its cocoon just like

the baby bird hatches from its egg.

17. Sequence On the diagram below, list the four stages of a butterfly's life cycle in the correct order.

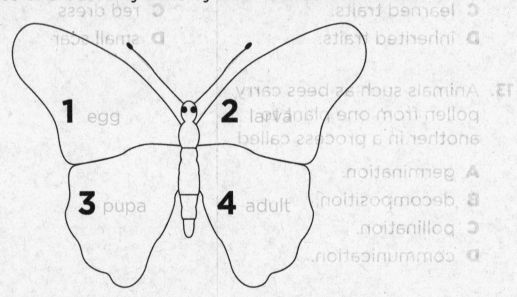

1 egg   2

3 pupa   4 adult

18. Predict What changes does a seed go through before it becomes a flower?

First the seed grows into a seedling. Given the proper food,

water, temperature, and sunlight, the seedling will grow into

a plant that produces flowers.

Answer the following questions.

19. A boy named Sam has brown, curly hair and blue eyes. Sam has a scar on his left knee. Sam can play the piano and ride a unicycle. List Sam's inherited traits, and then list Sam's learned traits.

_____

_____

20. How can the wind and animals help a plant to reproduce?

_____

_____

21. **Critical Thinking** How would you describe an environmental trait? List one example in your description.

_____

_____

_____

22. **Thinking Like a Scientist** Ferns reproduce without making seeds. Is a fern likely to grow in a location with a large population of pollinators such as bees? Explain your answer.

_____

_____

Comparing Life Cycles

Objective: Student will make a poster to compare the life cycles of two different animals.

Materials

- poster board
- markers or colored pencils
- old magazines and newspapers (for cutting out pictures)
- glue
- scissors
- reference materials (such as encyclopedias)

Scoring Rubric

4 points Student shows a good representation of all of the life cycle stages for both animals. Student clearly identifies all of the life cycle stages in correct sequential order for both animals. Student clearly explains his or her answers to the questions in Analyze the Results. Answers to all of the questions are correct.

3 points Student shows a good representation of most of the life cycle stages for both animals. Student clearly identifies most of the life cycle stages in correct sequential order for both animals. Student's answers to the questions in Analyze the Results are mostly correct.

2 points Student's representation of the life cycle stages for one or both animals is incomplete or out of sequential order. Student's answers to the questions in Analyze the Results are mostly inaccurate or incomplete.

1 point Student identifies some but not all of the life cycle stages of both animals. Student does not answer the questions in Analyze the Results.

Life Cycles

Make a Poster

Choose one animal from two of the following groups: amphibians, birds, insects, mammals, reptiles. Use the library or media center to find information about these animals. Draw or cut out pictures from magazines or newspapers of the animals to show the different stages of their life cycles. Then make a poster to display the life cycle of each of the animals. Be sure that your poster correctly identifies each phase of the animal's life cycle in the correct sequential order.

Analyze the Results

1. What are the differences in life cycles between the two animals? What are the similarities?

   Answers will vary depending on the animals selected.

   _____

   _____

   _____

2. Which animal would be better adapted to the area in which you live? Why?

   Answers will vary depending on the animal selected.

   _____

   _____

   _____

Living Things in Ecosystems

Write the word or words that best complete each sentence in the spaces below. Words may be used only once.

adaptation	consumer	mimicry	soil
camouflage	food chain	nocturnal	
climate	food web	producers	

1. Plants and algae are _____producers_____.

2. Bits of rock and humus form ____soil____ .

3. A giraffe's long neck is a(n) ____adaptation____ that helps it to survive in its environment.

4. Most desert animals are ____nocturnal____ and hunt at night.

5. The ____climate____ in most tropical rain forests is warm and wet.

6. Food chains can be linked together to form a(n) ____food web____ .

7. An organism that eats another organism is called a(n) ____consumer____ .

8. Turtles are higher on the ____food chain____ than insects.

9. Some animals use ____camouflage____ to blend into their environment.

10. When an organism imitates the color of another organism it uses ____mimicry____ .

© Macmillan/McGraw-Hill

Circle the letter of the best answer for each question.

11. What is the largest ocean in the world?

   A Atlantic Ocean

   B Pacific Ocean

   C Indian Ocean

   D Arctic Ocean

12. What is the thick layer of fat under walruses' skin?

   A blubber

   B fur

   C camouflage

   D flippers

13. Animals that hibernate

   A make their own food.

   B never sleep.

   C go into a deep sleep.

   D eat mostly plants.

14. Why do most desert animals come out at night and sleep during the day?

   A It is warm at night.

   B There are no predators at night.

   C It is cooler at night.

   D It is too windy during the day.

15. What is the name for organisms that break down dead plants and animals?

   A herbivores

   B producers

   C consumers

   D decomposers

Answer the following questions.

16. **Infer** Why does a rain forest have more types of organisms than a desert?

A rain forest has more sources of food for plant and animal survival, more rain, and moderate temperatures.

17. **Define** Explain how camouflage helps an animal stay safe in the ocean.

Camouflage helps an animal blend in with the colors and shapes in the ocean, making it harder to be seen or caught.

18. **Interpret Data** Look at the food chain below. Classify each organism as either a producer, a consumer, or a decomposer and tell why.

worm → algae → snail → fish → human

The worm is a decomposer because it feeds on dead plants and animals; algae is a producer because it makes its own food from the sun; a snail is a consumer because it feeds on the algae; a fish is a consumer because it feeds on the snail and the algae; a human is a consumer because it feeds on the fish.

19. **Infer** What is the main reason desert plants grow so far apart?

If they grew too close together, there would not be enough water or nutrients for all of them.

Answer the following questions.

20. Describe three adaptations that help animals survive in different climates.

Student's answer must include three of the following:

Camouflage: helps prey blend into their surroundings.

Mimicry: an animal will imitate another in color or shape to stay safe.

Hibernation: sleeping to protect itself from cold climates.

Blubber: a layer of fat keeps an animal warm in a cold climate.

Migration: an animal will move locations to look for food.

21. Critical Thinking What would happen if a large section of wetlands was eliminated to make room for a new shopping mall?

Possible answer: Flooding could occur. There would be a

depletion of oxygen in the area from the lack of plant life.

Erosion may occur. Animals would be displaced or killed.

22. Thinking Like a Scientist Imagine that you live in a rain forest and a friend has sent you a cactus as a gift. What would you need to do to keep the cactus alive in the rain forest?

Control its climate, keep the cactus's soil dry, and possibly

use a heat lamp to keep the temperature high.

Name _____   Date _____

Circle the letter of the best answer for each question.

1. How would you <u>best</u> describe a community in an ecosystem?

   A a single type of organism

   B only animals

   (C) all organisms

   D only the plants

2. How is a population of an ecosystem defined?

   (A) all the members of a single type of organism

   B all the members of two or more types of organisms

   C all the members of the animal families

   D all the members of the plant families

3. A producer

   A eats other organisms.

   (B) makes its own food.

   C breaks down dead organisms.

   D eats mostly plants.

4.

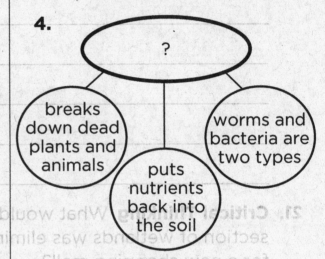

   Which belongs in the empty space?

   A Predator

   B Consumer

   C Producer

   (D) Decomposer

Critical Thinking How could a natural event, such as an earthquake or hurricane, change an ecosystem?

It could damage or destroy the plant life of an ecosystem. Animals

would not have enough to eat and could lose their homes.

Name _____ Date _____

Circle the letter of the best answer for each question.

1. What is a mixture of rocks and humus called?

   A soil

   B tundra

   C sand

   D bedrock

2. Which ecosystem has a very dry climate?

   A rain forest

   B desert

   C tundra

   D temperate forest

3. A tropical rain forest

   A is a dry ecosystem.

   B has the most kinds of living things.

   C has four distinct seasons.

   D is where few plants survive.

4. What type of ecosystem is represented by the word web below?

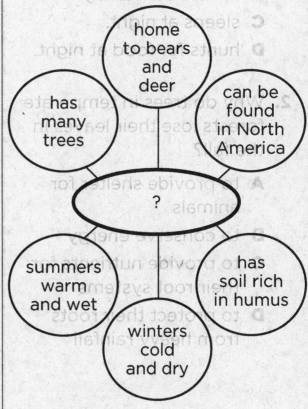

   A desert

   B temperate forest

   C wetland

   D rain forest

Critical Thinking How does a tropical rain forest differ from a temperate forest?

A tropical rain forest is warm all year long and gets a lot of rain,

whereas a temperate forest has four seasons and gets a moderate

amount of rain. There are also more living things in a tropical rain

forest than in any other ecosystem.

Circle the letter of the best answer for each question.

1. A nocturnal animal

   A only lives in deserts.

   B moves constantly.

   C sleeps at night.

   D hunts for food at night.

2. Why do trees in temperate forests lose their leaves in the fall?

   A to provide shelter for animals

   B to conserve energy

   C to provide nutrients for their root systems

   D to protect their roots from heavy rainfall

3. Air bladders help organisms

   A breathe.

   B take in food.

   C float.

   D gain energy.

4. What belongs in the empty circle below?

   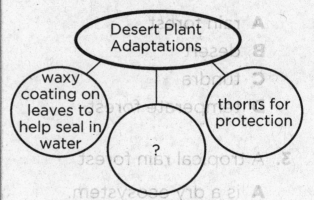

   A little or no roots

   B large leaves to catch sunlight

   C grooves and drip tips in leaves

   D thick stems to store water

Critical Thinking What might happen to a dormouse if it did not hibernate during the winter?

The dormouse could die because it might not be able to find

enough food and water, or it might have to migrate to a different

environment in order to survive.

Living Things in Ecosystems

Write the word or words that best complete each sentence in the spaces below. Words may be used only once.

adaptation	decomposers	food web	wetlands
climate	ecosystem	forests	
consumers	food chain	migrate	

1. The typical weather conditions of an area over
   time are called its ___climate___ .

2. A ___food chain___ shows how energy passes from
   one organism to another in an ecosystem.

3. A ___food web___ shows how a group of food chains
   are linked together.

4. Marshes, swamps, and bogs are all examples of ___wetlands___.

5. Temperate ___forests___ have four distinct seasons.

6. Animals are ___consumers___ because they cannot
   make their own food.

7. Worms are ___decomposers___ that break down dead
   plant and animal materials.

8. Some birds ___migrate___ south for the winter.

9. A special characteristic that helps a living thing
   survive in its environment is called a(n) ___adaptation___ .

10. Living and nonliving things that share an
    environment are part of a(n) ___ecosystem___ .

Circle the letter of the best answer for each question.

11. What adaptation helps jackrabbits stay cool in the desert?

   A small eyes
   B soft fur
   C long legs
   (D) large ears

12. What do most desert animals do at night?

   A They sleep.
   B They burrow.
   (C) They hunt for food.
   D They become scared.

13. Which of the following statements about ocean algae is false?

   A Some algae only live in shallow water.
   B Algae use sunlight to make food.
   C Some algae can live without roots.
   (D) Algae can only survive in deep, dark water.

14. Where do most ocean organisms live?

   A on islands
   B deep waters
   C shallow waters
   D the ocean's surface

15. The soil in a tropical rain forest is

   (A) not very rich in nutrients.
   B rich in nutrients.
   C dry and cold.
   D mostly sand.

Answer the following questions.

16. **Define** Explain how mimicry can help an animal stay safe in the forest.

An animal pretends to have the same characteristics as an

animal that its predator does not like to eat.

17. **Interpret Data**
    Look at the food
    web to the right.
    Where is this food web
    most likely found?
    How can you tell?

It is most likely found in a pond. I can tell because of the kinds

of plants and animals in the food web.

18. **Infer** Why do you think nutrients can be found only in the top layer of soil in a rain forest?

Nutrients are absorbed very quickly and do not have the

chance to penetrate to the underlayers of the forest floor.

19. **Infer** Why do most desert plants have such far-reaching roots?

The water in desert soil is deeper than it is in other

environments. There is no humus in desert soil, only sand. So

the water is contained far beneath the surface.

Answer the following questions.

20. Name an adaptation that helps rain forest plants survive in their environment. Explain why this adaptation helps the plants.

Rain forest plants have large leaves that help them survive

on the forest floor where little sunlight reaches them. Some

leaves have special tips to help rainwater flow off them and

prevent damage.

21. Critical Thinking If a rain forest on the other side of the world was eliminated, would it have any effect on life in the United States? Why or why not?

Answers will vary but may include: It would have an effect

on life here in the United States because rain forests produce

much of the world's oxygen. Rain forests also produce many

medicines that are used in the United States.

22. Thinking Like a Scientist Imagine that you are a scientist studying migrating whales. Why do you think some whales migrate such long distances?

Answers will vary but may include: Some whales may migrate

long distances to find food or warmer waters.

Identify and Illustrate Ecosystems

Materials

• poster board
• markers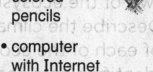
• colored pencils
• computer with Internet access

Objective: Students will illustrate a world map that shows two separate ecosystems. Students will identify plants and animals that live in each of the chosen ecosystems. Students will display their ecosystem maps throughout the classroom. Students will compare their maps with other students' maps so they can discuss the similarities and differences among the different ecosystems.

Scoring Rubric

4 points Student clearly labels and describes two ecosystems on the world map and includes descriptions of animals and plants found within these ecosystems. Student clearly explains his or her answers to the questions in Analyze the Results.

3 points Student clearly labels two ecosystems and includes descriptions of plants and animals within those ecosystems, but does not describe the climate and/or environment of each ecosystem. Student answers the questions in Analyze the Results. Answers to the questions are partially correct.

2 points Student identifies and describes only one ecosystem. Student provides limited descriptions of the plants and animals in the ecosystem. Student attempts to answer the questions in Analyze the Results. Answers to the questions are incorrect.

1 point Student describes and/or labels one ecosystem, but does not provide details about climate, environment, plants, or animals. Student does not answer the questions, or the answers are incorrect.

Name _____ Date _____

Ecosystems of the World

Communicate

Use the Internet to find the location of two of the ecosystems listed below. Describe the climate and environment of each of the ecosystems you select. Identify plants and animals that live in each of the ecosystems. Using a world map, mark and label where your ecosystems are located.

Internet Tips to Stay Smart and Safe

✔ DO visit Web sites that will help you with your project.
✔ DO ask a teacher for help if you get lost.
✘ DO NOT talk to strangers on the Internet.

Ecosystems

arctic tundra	ocean	tropical rain forest
desert	temperate forest	wetland

Analyze the Results

1. Of the two ecosystems, which one covers more area on your map?

   Answers will vary depending on the ecosystem chosen.

   _____

   _____

2. Choose one animal and one plant that you listed and describe the adaptations that each uses to survive.

   Answers will vary depending on the plant and animal

   selected.

   _____

© Macmillan/McGraw-Hill

Changes in Ecosystems

Write the word that best completes each sentence in the spaces below. Words may be used only once.

community	endangered	fossils	reuse
competition	extinct	pollution	
drought	flood	resources	

1. Food, water, and air are all considered _____ .

2. To _____ means to use something again.

3. When there is no rain for a long time, a(n) _____ can occur.

4. All living things in an ecosystem make up a _____ .

5. The remains of plants or animals that lived long ago are called _____ .

6. The struggle among living things for survival is _____ .

7. A living thing is _____ when there are no more of its kind alive.

8. Harmful material that damages the air, water, or land is called _____ .

9. A natural disaster that happens when dry land becomes covered by water is a(n) _____ .

10. The Saharan cypress is _____ because there are few of these trees left.

Circle the letter of the best answer for each question.

11. Which of the following live in the soil and help the environment?

    A spiders and their webs

    B worms, bacteria, and fungi

    C turtles and their eggs

    D squirrels and their acorns

12. What causes more changes to the environment than anything else?

    A people

    B migrating animals

    C floods

    D drought

13. Which of the following statements is false?

    A People often cause permanent changes to the environment.

    B All plants and animals have survived in their ecosystems.

    C Scientists study fossils to learn about the past.

    D Not all changes are bad for the environment.

14. What does a fossilized sharp tooth tell scientists about the animal it came from?

    A The animal was probably a meat eater.

    B The animal was probably a plant eater.

    C The animal was very aggressive.

    D The animal was very young.

15. Which of the following is <u>not</u> a reason that an organism can become endangered?

    A People have destroyed its environment.

    B The organism's climate has changed.

    C The organism migrates in the winter.

    D Hunters have killed too many of the organism.

Answer the following questions.

16. **Interpret Data** Study the information in the pie chart. How much more trash in the United States is sent to landfills than recycled?

Trash in the United States

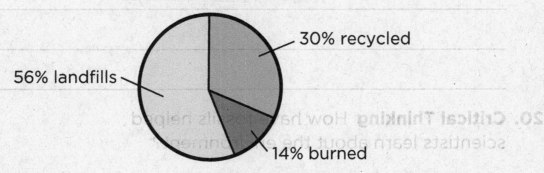

30% recycled

56% landfills

14% burned

26% more trash is sent to landfills than is recycled.

17. **Infer** What might happen to plant and animal populations in the desert if the climate changes and becomes wet and cool?

The plants would have to adapt to the cooler, wetter

conditions. Plants unable to adapt could become extinct.

Animals would either have to adapt to the new environment,

or migrate to more suitable new regions.

18. **Communicate** Tell why it is important to reduce, reuse, and recycle.

Answers will vary but may include: It is important to reduce,

reuse, and recycle in order to conserve resources and reduce

waste and pollution. If people stop doing this, resources will

run out and landfills will overflow.

Answer the following questions.

19. What happens to animals that cannot adapt to environmental changes?

They may have to move to find new sources of food and shelter. If they are not able to move, they could become endangered or even become extinct.

20. **Critical Thinking** How have fossils helped scientists learn about the environment?

Fossils can tell an animal's shape and size, and what it ate. Fossils can also provide information about Earth during the animal's life; for example, the fossil can tell whether the climate was hot or cold or dry or wet.

21. **Thinking Like a Scientist** Write a short scientific report explaining the reasons why a large tree in your local park should not be cut down.

Answers will vary but may include: If the tree is cut down, many animals may be left without a home. Tree roots help to keep soil in place. By cutting down the tree, the soil may get washed or blown away. The tree also provides oxygen and shade.

Circle the letter of the best answer for each question.

1. When living things battle each other for resources, it is called

   A producing.

   B recycling.

   C cooperation.

   D competition.

2. People harm the environment

   A by recycling.

   B by draining wetlands.

   C by reducing waste.

   D by reusing products.

3. Which of the following causes the <u>most</u> change to the environment?

   A animals

   B plants

   C people

   D storms

4. What is being represented by the word web below?

   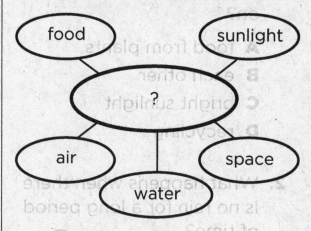

   A Resources

   B Organisms

   C Competition

   D Pollution

Critical Thinking What might happen if too many new towns are built in an environment?

It would hurt natural environments by draining wetlands and

removing forests. Many plants and animals are then left without

homes. Soil may also wash away, causing erosion.

_____

Circle the letter of the best answer for each question.

1. In an ecosystem, what do all living things depend on?

   A food from plants

   B each other

   C bright sunlight

   D recycling

2. What happens when there is no rain for a long period of time?

   A a drought can occur

   B a flood can occur

   C a watering hole may form

   D mildew can occur

3. What do plants and animals have that help them survive in their environment?

   A adaptations

   B resistance to fires

   C ability to store water

   D prey

4. Study the chart. Which animal most likely needs to be protected?

Animal	Status
dog	common
dodo bird	extinct
marine turtle	endangered
house cat	common

   A dog

   B dodo bird

   C marine turtle

   D house cat

Critical Thinking How can a drought lead to a wildfire?

Trees and other plants become dry during a drought. If lightning strikes a dry area, a wildfire can start very quickly.

Circle the letter of the best answer for each question.

1. According to some scientists, about how many animal species become extinct each day?

   A 1

   B 10

   C 100

   D 1,000

2. What modern-day animal is the woolly mammoth most similar to?

   A a turtle

   B an elephant

   C a tiger

   D a buffalo

3. Why did the St. Helen Olive tree become extinct?

   A It was destroyed by forest fires.

   B It was cut down.

   C It was destroyed by disease and dry weather.

   D It was washed away in a flood.

4. Which animal was most likely a meat-eater?

Animal	Feature
saber-toothed cat	long, sharp teeth
woolly mammoth	flat teeth
pterodactyl	long wingspan
triceratops	leathery skin

   A saber-toothed cat

   B woolly mammoth

   C pterodactyl

   D triceratops

Critical Thinking What can a fossil's depth in the ground tell us?

Fossils found deeper in the ground are older than fossils found

closer to the surface.

Name _____   Date _____

Changes in Ecosystems

Write the word that best completes each sentence in the spaces below. Words may be used only once.

competition	flood	population	resources
endangered	fossils	recycle	
extinct	pollution	reduce	

1. One way to help the environment is to
   _____reduce_____ our use of certain products.

2. Two animals that fight for the same food source is
   an example of ____competition____ .

3. The St. Helena Olive tree became ____extinct____
   because of disease and dry weather.

4. When harmful materials damage the air, water, or
   land, this is called ____pollution____ .

5. Sunlight and air are ____resources____ that help an
   organism survive.

6. The remains of organisms that lived long ago are
   ____fossils____ .

7. When only a few of an organism are left, it is ____endangered____ .

8. All the eagles in an ecosystem make up a(n) ____population____ .

9. A person can ____recycle____ an old tire by turning
   it into a tire swing.

10. Heavy rains and other storms can cause a(n) ____flood____ .

Circle the letter of the best answer for each question.

11. What could happen if people introduce a new animal into an environment?

   A the environment will not be affected

   B the animal uses up resources

   C other animals will create pollution

   D all plants and animals will live longer

12. An example of something that is harmful to the environment is

   A a bird that builds a nest on a rooftop.

   B a lake filled with fresh water.

   C an oil spill in the ocean.

   D a plant that decomposes.

13. Which of the following cannot be recycled?

   A plastics

   B metals

   C glass

   D coal

14. Which of the following is a reason why pandas are endangered?

   A people have destroyed the forests where they live

   B the water sources in their environment have dried up

   C they are no longer able to migrate in the winter

   D hunters kill them for their fur

15. What do fish fossils found on land tell us?

   A that there were land-roaming fish

   B that part of the land was once covered in water

   C that people once fished on the beach

   D that many different animals fed on fish

Name _____   Date _____

Answer the following questions.

16. **Communicate** Explain what your school can do to help the environment.

Answers will vary but may include: recycle paper, glass, and plastic.

_____

_____

17. **Interpret Data** Study the information in the bar graph. Which material is recycled the most? How much of that material is recycled monthly?

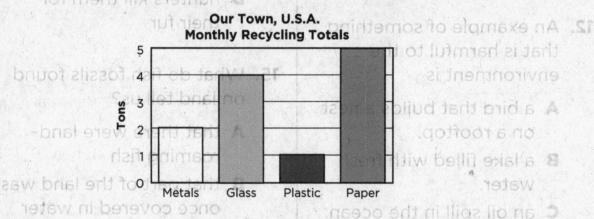

Our Town, U.S.A.
Monthly Recycling Totals

Paper is recycled the most. Five tons of paper are recycled each month.

_____

18. **Infer** If a predator in an environment died out, what do you think would happen to the predator's prey?

Answers will vary but may include: The predator's prey

would multiply greatly and use up all the resources. The

environment would no longer be balanced.

Answer the following questions.

19. What are the causes and effects of wildfires?

Wildfires may be caused by lightning. Droughts may lead to

wildfires because dry plants burn easily. Wildfires can destroy

habitats and pollute the air with smoke.

20. Critical Thinking How can you tell what an animal ate by examining its fossil?

The best way to tell what an animal ate is by examining the

shape of its teeth. If the animal had sharp teeth, it probably

was a predator and ate meat. If the animal had flat teeth, it

probably ate plants.

21. Thinking Like a Scientist Your town has some open space. Some people want to build a new shopping center but you think more trees should be planted there. Write a short report supporting this idea.

Answers will vary but may include: We should plant trees

instead of building a shopping center. Trees provide oxygen,

food, and shelter for animals. The trees can also help prevent

erosion.

© Macmillan/McGraw-Hill

Recycling Brochure

Make a Recycling Brochure

Objective: Students will make a brochure describing and explaining the importance of recycling.

Scoring Rubric

4 points   Student accurately makes a brochure describing the importance of recycling. Student clearly draws and labels items that can be recycled and those items that cannot be recycled. Student clearly explains his or her answers to the questions in Analyze the Results.

3 points   Student makes a brochure describing the importance of recycling. Student draws and labels multiple items that can be recycled and several that cannot be recycled. Student answers the questions in Analyze the Results. Answers to the questions are partially correct.

2 points   Student makes a brochure about recycling. Student labels one or two items that can be recycled or not recycled. Student attempts to answer the questions in Analyze the Results. Answers to the questions are incorrect.

1 point   Student draws a picture about recycling but does not make a brochure. Student does not use descriptive words to show the importance of recycling. Student may list or draw one item that can be recycled or cannot be recycled but does not provide multiple examples. Student does not answer the questions.

Materials

- construction paper
- markers or crayons
- pencil
- colored pencils
- magazines
- newspapers

© Macmillan/McGraw-Hill

Recycling Brochure

Communicate

Use the materials your teacher has given you to make
a brochure that explains the importance of recycling.
Design the brochure so that people know exactly what
they can and cannot recycle. Explain in your brochure
the 3 Rs (reduce, reuse, and recycle). Use descriptive
words that tell why recycling can help save our
environment.

Analyze the Results

1. Which words in your brochure express why
    recycling is important?

    Answers will vary but may include: "Save our environment.

    Recycle newspapers, cans, and bottles. Reduce, Reuse,

    Recycle. Our future depends on you."

2. Explain how recycling can cut down on pollution.

    By recycling, you are cutting down on the waste materials

    that are put back into our environment.

3. What message do you want your community
    to learn?

    Answers will vary but may include: Recycling helps keep the

    environment clean and healthy.

Earth Changes

Write the word that best completes each sentence in the spaces below. Words may be used only once.

core	erosion	magma	weathering
crust	landform	mantle	
earthquake	lava	volcano	

1. The sudden movement of plates on Earth's crust is

   called a(n) _____earthquake_____ .

2. Melted rock that flows out of a volcano is _____lava_____ .

3. The deepest and hottest layer of Earth, located at

   its center, is called the _____core_____ .

4. The slow process of breaking down rocks into

   smaller pieces is called _____weathering_____ .

5. A natural feature of land, is a(n) _____landform_____ .

6. Melted rock inside Earth's mantle and crust is _____magma_____ .

7. The movement of weathered rock, sometimes

   caused by moving water or wind is called _____erosion_____ .

8. The layer below Earth's crust is the _____mantle_____ .

9. A mountain that builds up around an opening in

   Earth's crust is called a(n) _____volcano_____ .

10. Earth's outermost layer on which the continents

    are located is called Earth's _____crust_____ .

Circle the letter of the best answer for each question.

11. What are the seven great areas of land on Earth called?

    A countries

    B continents

    C landforms

    D volcanoes

12. Which of the following can suddenly change the land on Earth?

    A weathering

    B glacier

    C earthquake

    D erosion

13. All of the following may cause erosion <u>except</u>

    A sunlight.

    B water.

    C wind.

    D glaciers.

14. Which of the following statements is <u>false</u>?

    A Almost half of Earth is covered by water.

    B Oceans are large bodies of salt water.

    C A trench is a canyon on the ocean floor.

    D Earth's deepest layer is called the core.

15. An epic movement of rocks and soil down a hill is called a(n)

    A earthquake.

    B volcano.

    C tsunami.

    D landslide.

Answer the following questions.

You are going on a trip. Use the map below to help you plan for the landforms you will encounter along the way.

Map key:
≋ Mountains
▦ Plains
⅗ Canyon
⤳ Valley
⥈ River

16. **Interpret Data** Look at the map key. How can reading the key help you plan your trip?

Answers will vary but may include the following: The map key tells me what kinds of landforms I will encounter, how to travel, and what clothing to bring on the trip.

17. **Infer** How does the west side of the map differ from the east side? What landform separates the east from the west?

The west side of the map consists of a plains region and the east side consists of a mountainous region. The east side and west side are separated by a canyon and river.

18. **Define** How would you explain a canyon to someone unfamiliar with the term?

A canyon is a deep, narrow valley with steep sides. Canyons often have rivers flowing through them. The rivers often cause erosion, which is what carves out the canyon.

Answer the following questions.

Label Earth's layers.

19. _____

20. _____

21. _____

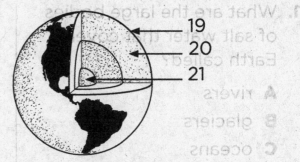

19
20
21

22. **Critical Thinking** Discuss the sudden changes and slow changes a hurricane could cause to Earth's surface.

Sudden changes—flooding; changes to shoreline; plants and trees uprooted, thereby displacing soil; Slow changes— weathering from rain and heavy winds; erosion from winds and flooding.

23. **Thinking Like a Scientist** You are asked to research the Mariana Trench in the Pacific Ocean. What landform might you use to aid in your research? Why?

Answers will vary but may include the following: I would research the Grand Canyon because an ocean trench is very much like a canyon on land. The Grand Canyon would be a good example because it is a very large canyon just as the Mariana Trench is a very large ocean trench.

Circle the letter of the best answer for each question.

1. What are the large bodies of salt water that cover Earth called?

   A rivers
   B glaciers
   C oceans
   D lakes

2. North America is an example of a

   A country.
   B continent.
   C landform.
   D state.

3. What is the name for the land that lies at the bottom of the ocean?

   A ocean floor
   B ocean plateau
   C ocean coast
   D ocean canyon

4. What lies under the ocean at the edge of a continent?

   A ocean volcano
   B continental divide
   C ocean currents
   D continental shelf

Critical Thinking How do Earth's landforms affect Earth's water supplies?

Earth's landforms affect how water flows and is stored. For

example, a mountain can cause water to run downhill, which helps

feed rivers and streams. If there were no mountains, then the

water would likely stay in one place. Indentations in the Earth's

crust can cause water to pool, which helps form lakes and ponds.

_____

_____

Circle the letter of the best answer for each question.

1. What occurs when water rises over the banks or sides of a river?

A a landslide

B an earthquake

C a flood

D a hurricane

2. When magma flows onto land, it is called

A crust.

B lava.

C ash.

D mantle.

3. What happens when hardened lava builds up over a long period of time?

A mountains form

B rivers form

C an earthquake occurs

D a flood occurs

4. Which of the following forces can cause a landslide?

A drought

B windstorms

C magnetic force

D gravity

Critical Thinking What affect would a heavy snowfall have on rivers and streams?

Heavy snowfalls would cause rivers and streams to rise when the

snow melts. The rising water levels in rivers and streams could

cause flooding and erosion. The melting snow and ice could also

cause landslides on mountains.

Circle the letter of the best answer for each question.

1. Which of the following is <u>not</u> known to cause weathering?

   A rain

   B ice

   C light

   D wind

2. What is a mass of ice that moves slowly across land called?

   A a glacier

   B an iceberg

   C an ice cap

   D an avalanche

3. How can gravity cause erosion?

   A it causes flooding

   B it heats the rocks

   C it pulls rocks downhill

   D it wears away the land

4. How can planting trees help the land?

   A Trees add carbon dioxide to the air.

   B Trees help prevent erosion.

   C Trees provide more timber for houses.

   D Trees drain nutrients from the soil.

Critical Thinking How can ice cause weathering?

When water freezes and forms ice, it expands. If water gets into

rocks and freezes, it will expand and cause the rocks to crack

over time. As the rocks crack, they will break away and cause

weathering to occur.

Earth Changes

Write the word that best completes each sentence in the spaces below. Words may be used only once.

core	erosion	magma	weathering
crust	landforms	mantle	
earthquake	lava	volcano	

1. The center of Earth is the _____ core _____.

2. Breaking down rocks into smaller pieces is a slow process called _____ weathering _____ .

3. Rock that is melted and flows onto Earth's surface is called _____ lava _____ .

4. A(n) _____ earthquake _____ is the sudden movement of rocks on Earth's crust.

5. Earth's outermost layer is called the _____ crust _____ .

6. The wearing away of rocks is _____ erosion _____ called.

7. The layer below Earth's crust is the _____ mantle _____ .

8. Melted rock that is found beneath Earth's mantle and crust is called _____ magma _____ .

9. Land features, such as plains and valleys, are called _____ landforms _____ .

10. A(n) _____ volcano _____ is a mountain that builds up around an opening in Earth's crust.

© Macmillan/McGraw-Hill

Circle the letter of the best answer for each question.

11. Which of the following causes slow changes to Earth's surface?

   A tsunami

   B earthquake

   C hurricane

   (D) glacier

12. Gravity can pull rocks and soil rapidly down a hillside, causing a(n)

   A tsunami.

   B earthquake.

   (C) landslide.

   D volcano.

13. All of the following statements are true <u>except</u>

   A Earth's deepest layer is called the core.

   B a trench is a canyon on the ocean floor.

   C oceans are large bodies of salt water.

   (D) more than half of Earth is covered by land.

14. Which of the following does <u>not</u> cause erosion?

   A water

   (B) sunlight

   C glaciers

   D wind

15. There are seven large land areas on Earth called

   (A) continents.

   B volcanoes.

   C countries.

   D landforms.

© Macmillan/McGraw-Hill

Answer the following questions.

Use the pie chart below to help you plan for the landforms you will encounter on your trip.

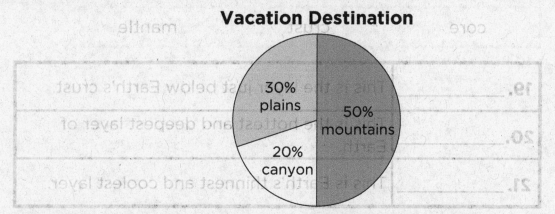

Vacation Destination

30% plains

50% mountains

20% canyon

16. **Infer** Should you bring hiking shoes on your trip? Explain why or why not.

I should bring hiking shoes because 50% of the land is

mountainous and 20% is a canyon.

17. **Define** How would you explain what a mountain is to someone who is unfamiliar with the term?

A mountain is the tallest landform. It often has steep sides

and a pointed top.

18. **Interpret Data** What three types of landforms will you encounter on your trip? Rank them from the type you will see the greatest number of, to the type you will see the least number of.

1. _mountains_

2. _plains_

3. _canyon_

Answer the following questions.

19.–21. Complete the chart with words from the box below.

core	crust	mantle

19. ___mantle___	This is the layer just below Earth's crust.
20. ___core___	This is the hottest and deepest layer of Earth.
21. ___crust___	This is Earth's thinnest and coolest layer.

22. Critical Thinking Volcanoes can both destroy and create landforms. Explain how this can be true.

When lava explodes out of a volcano, landforms like

mountains can crumble under the pressure. However, if lava

from a volcano slowly oozes out, it can harden and make the

mountain bigger.

23. Thinking Like a Scientist If you were to compare the Grand Canyon to a landform that is located on the ocean floor, what landform would you compare it to and why?

Answers will vary but may include the following: I would

compare the Grand Canyon to the Mariana Trench because:

a) an ocean trench can be best compared to a canyon on

land; b) the Grand Canyon is the largest canyon in the United

States and the Mariana Trench is the largest ocean trench.

Letter to the Editor

Materials

- paper
- markers
- pencils

Objective: Student will write a letter to a newspaper editor explaining why clearing a wooded area would be harmful to the environment. Student's statements will be supported by facts he or she has learned.

Scoring Rubric

4 points Student correctly describes the sudden and long-term effects that building a shopping center would have on the environment and gives clear examples of each. Student clearly and correctly explains his or her answers to the questions in Analyze the Results.

3 points Student shows a good understanding of the sudden and long-term effects and gives examples of each. Examples may not be well elaborated. Answers to the questions in Analyze the Results are mostly correct, but contain a few errors.

2 points Student's understanding of the sudden and long-term effects are somewhat inaccurate. Student may offer only long-term or only sudden effects, but not both. Student's answers to the questions are mostly inaccurate or incomplete.

1 point Student misunderstands the concepts of sudden and long-term effects on the environment. Student does not answer the questions in Analyze the Results.

Letter to the Editor

Communicate

A local developer has submitted a proposal to the city council to clear a wooded area in your community to build a shopping center. Write a letter to the newspaper editor that opposes the construction of the new shopping center. Be sure to state the reasons why a new shopping center would be harmful to the environment. Discuss the sudden and long-term effects that removing the natural resources would have on the environment and the people living there.

Analyze the Results

1. Name some of the sudden effects that clearing the land may cause to the environment.

    Answers will vary but should include the following: Depleting the land of plants and trees will cause the soil to become displaced. When it rains, there will be nothing to catch the water and the run-off may cause flooding and/or landslides.

2. What long-term effects would clearing the land have on the environment?

    Answers will vary but should include the following: Over time the depletion of trees and other natural resources such as rocks, will cause erosion.

© Macmillan/McGraw-Hill

Using Earth's Resources

Write the word or words that best completes each sentence in the spaces below. Words may be used only once.

conservation	nonrenewable resource	sedimentary rock
igneous rock	pollution	solar energy
mineral	renewable resource	
natural resource	sediment	

1. A resource that can be used over and over is a(n) __renewable resource__ .

2. Tiny bits of rock or dead plant and animal matter is called __sediment__ .

3. A rock that is formed from layers of sediment is called a(n) __sedimentary rock__ .

4. Harmful things in the air cause __pollution__ .

5. Melted rock cools and hardens, forming __igneous rock__ .

6. Making resources last longer is called __conservation__ .

7. A material on Earth that is necessary or useful to people is a(n) __natural resource__ .

8. Energy from the sun is called __solar energy__ .

9. Salt is a type of __mineral__ .

10. A resource that takes millions of years to form and cannot be easily replaced is a(n) __nonrenewable resource__ .

Circle the letter of the best answer for each question.

11. What is another name for a valuable mineral such as a diamond or a ruby?

A gem

B coal

C topaz

D igneous rock

12. What is water that is held in rocks below ground called?

A groundwater

B aqueducts

C fresh water

D reservoirs

13. Which mineral property is identified by the color of the powder left behind when it is rubbed across a surface?

A hardness

B streak

C color

D luster

14. What are the impressions that living things leave behind in the mud called?

A casts

B remains

C fossils

D imprints

15. Which part of the soil works like a sponge?

A bedrock

B humus

C subsoil

D sand

© Macmillan/McGraw-Hill

Answer the following questions.

Use the diagram to the right to answer the following question.

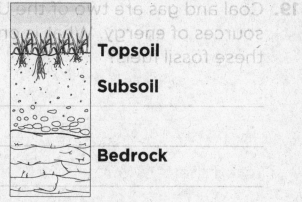

Topsoil

Subsoil

Bedrock

_____

_____

16. Interpret Data Which level of soil is rocky and would be the most difficult to reach? Why?

Bedrock is rocky and is the most difficult to reach because

it is the deepest layer of soil. In order to reach it, one would

have to dig through both the topsoil and the subsoil.

17. Infer Why is it important for farmers to prevent erosion?

Farmers prevent erosion because the topsoil contains the

nutrients that crops need to grow. It takes a long time for

topsoil to form, so farmers do not want it to wash away.

18. Classify Use the following vocabulary terms to classify each rock.

| igneous rock | metamorphic rock | sedimentary rock |

granite _____igneous rock_____   sandstone _____sedimentary rock_____

slate _____metamorphic rock_____

Answer the following questions.

19. Coal and gas are two of the United States's main sources of energy. What is one problem with using these fossil fuels?

Answers will vary but may include the following: Coal and gas

are both nonrenewable resources. Both fuels create pollution

when they are burned.

20. Many people are considering alternative forms of energy. What alternative form of energy do you think would be best for the environment? Why?

Answers will vary but may include the following: Solar or

wind energy, because they are renewable resources.

21. **Critical Thinking** Some animals live in soil. How can animals benefit the soil where they live?

By moving around, animals can break up the soil. Animals'

homes in the soil can help air and water get into it. Some

animals, like earthworms, put nutrients into the soil.

22. **Thinking Like a Scientist** You live in an area with very little water. There is a water source several miles away. Describe one way you could move the water from its source to where it is needed.

Answers will vary but may include information about an

aqueduct, a canal, ditch, or pipe that carries water to where it

is needed.

© Macmillan/McGraw-Hill

Circle the letter of the best answer for each question.

1. How does sedimentary rock form?

   A from volcanic ash that becomes compressed over time

   B from tiny bits of weathered rock, animal remains, or plants that pile up in layers

   C by heating and squeezing rocks and minerals together

   D by cooling and hardening melted rock

2. Each of the following is a property of minerals <u>except</u>

   A color.

   B streak.

   C luster.

   D size.

3. What is meant by a mineral's luster?

   A how difficult it is to scratch

   B how light reflects off of it

   C the color of the powder left behind

   D how it is formed

4. What is the term used to describe any kind of rock that is changed by heating and squeezing?

   A metamorphic rock

   B sedimentary rock

   C igneous rock

   D obsidian rock

Critical Thinking Why should a student not use color alone to identify minerals?

Some minerals come in many colors and sometimes different

minerals are the same color.

_____

Circle the letter of the best answer for each question.

1. When minerals, weathered rocks, and other things are mixed together, they form

   A peat.

   B sand.

   C humus.

   D soil.

2. What is humus?

   A the sandy part of soil

   B roots that grow deep into the soil

   C bits of decayed plants and animals found in soil

   D the top layer of soil

3. What is the correct order of layers of soil from top to bottom?

   A bedrock, subsoil, topsoil

   B topsoil, subsoil, bedrock

   C topsoil, bedrock, subsoil

   D subsoil, topsoil, bedrock

4. What determines a soil's color?

   A how much sun it gets

   B how much water it gets

   C what types of materials are in it

   D what grows on it

Critical Thinking A student is growing plants that need soil that is neither too wet nor too dry. What would be a good soil to use? Why?

Loam would be a good soil in which to grow these plants because

loam is made up of a mixture of sand, silt, and clay. It is neither

too wet nor too dry.

Name _____ Date _____

Circle the letter of the best answer for each question.

1. What are imprints?

   A types of shells

   B types of fossils

   C types of minerals

   D types of plants

2. Which is one of Earth's main sources of renewable energy?

   A nuclear energy

   B electricity

   C solar energy

   D fossil fuel energy

3. All of the following are renewable resources <u>except</u>

   A natural gas.

   B plants.

   C water.

   D air.

4. All of the following are examples of fossil fuels <u>except</u>

   A coal.

   B natural gas.

   C water.

   D oil.

Critical Thinking Why is it important to conserve fossil fuels?

Fossil fuels are nonrenewable resources. There are limited

supplies of fossil fuels. If people do not conserve fossil fuels, then

they will run out and it takes millions of years to make more.

© Macmillan/McGraw-Hill

Circle the letter of the best answer for each question.

1. Where is most of Earth's water located?

   A rivers

   B lakes

   C oceans

   D glaciers

2. What do rivers, ponds, streams, glaciers, and icebergs all contain?

   A fresh water

   B salt water

   C drinking water

   D frozen water

3. What is an aqueduct?

   A a wall across a river

   B a place to store water

   C another name for groundwater

   D a pipe or ditch that carries water

4. What kind of water is pumped from wells?

   A ocean water

   B treated water

   C groundwater

   D river water

Critical Thinking Why is it important for cities to have water treatment plants?

Water treatment plants can trap and filter out harmful materials in

the water. Water treatment plants can also pump water to many

people in many different places so that even people who do not

live near a source of water can have water.

© Macmillan/McGraw-Hill

Using Earth's Resources

Write the word or words that best complete each sentence in the spaces below. Words may be used only once.

conservation	mineral	renewable resource
fossils	natural resource	sedimentary rock
fuel	nonrenewable resource	
igneous rock	pollution	

1. Sulfur is a type of _____ .

2. The burning of fossil fuels causes _____ .

3. Using resources wisely is called _____ .

4. Coal is an example of a(n) _____ that cannot be easily replaced.

5. Water is an example of a(n) _____ that can be used again and again.

6. Sandstone is an example of a _____ .

7. The remains or traces of something that lived long ago are _____ .

8. Soil is an example of a(n) _____ .

9. Magma cools slowly and then hardens into a(n) _____ .

10. Energy used in homes and in cars is called _____ .

© Macmillan/McGraw-Hill

Circle the letter of the best answer for each question.

11. What is groundwater?

A salt water from oceans

B water that falls to the ground when it rains

C water that is held in rocks below ground

D freshwater taken from reservoirs and aqueducts

12. Which mineral property is identified by how light bounces off it?

A luster

B hardness

C streak

D color

13. All of the following are examples of gems except

A diamond

B topaz

C slate

D ruby

14. Which part of the soil soaks up rainwater, keeps the soil moist, and adds nutrients?

A sand

B bedrock

C subsoil

D humus

15. How are imprints best described?

A bones and shells from long ago

B impressions that living things leave behind in the mud

C valleys carved out by glaciers

D minerals that seep into a mold and harden

Answer the following questions.

16. **Classify** Match each rock type with its example.

slate ____A____   **A** metamorphic rock

shale ____C____   **B** igneous rock

granite ____B____   **C** sedimentary rock

17. **Interpret Data** Choose from the box below to
complete the chart.

| subsoil | topsoil | bedrock |

Soil Layer	Description
topsoil	This layer contains the most humus and minerals and is dark in color.
subsoil	This layer contains less humus and is lighter in color.
bedrock	This layer is made of solid rock.

18. **Infer** What type of plant would grow best in sandy
soil? Why?

Answers will vary but may include the following: A cactus or

other plant that can grow well in dry soil. Sandy soil is made

up of small grains of sand, and it does not hold water as well

as other types of soil. A plant that does not need a lot of

water would grow best.

Answer the following questions.

19. What alternative form of energy would work best in the environment where you live? Why?

Answers will vary. Accept all reasonable responses.

_____

_____

20. List three ways to reduce the use of fossil fuels.

Answers may vary but may include the following: Drive more

fuel-efficient cars; use public transportation; walk or ride a

bicycle instead of driving; carpool.

21. **Critical Thinking** Why are earthworms important in soil development?

Answers will vary but may include: Earthworms burrow in the

soil. This helps the soil "breathe" by letting air and water in.

22. **Thinking Like a Scientist** The Hoover Dam is located on the borders of Nevada and Arizona. The Hoover Dam provides water to millions in Arizona, California, and Nevada. Why is it more important to provide water to these states than to rainy states like Washington or Oregon?

Arizona, California, and Nevada contain large desert areas.

The dam can store water and feed it to these areas as needed.

Washington and Oregon have large amounts of rainfall and

would likely have water stored in their own reservoirs.

_____

The Water We Use

Materials

• poster board

• markers or colored pencils

• reference materials (such as encyclopedias, Internet sites) – for ideas only

Objective: Student will create a diagram showing the path drinking water takes from the treatment plant to their faucet at home.

Scoring Rubric

4 points Student shows a good representation of each step in the water treatment process. Student clearly labels each step of the process. Steps should include all of the following: Water enters the plant; water passes through a strainer or filter; water sits in a settling tank so large particles can sink to the bottom; water is filtered through sand or gravel; water is chemically treated to kill harmful organisms; water is piped to homes. Student clearly explains his or her answers to the questions in Analyze the Results.

3 points Student shows a good representation of most of the steps in the water treatment process, leaving out one step or label. Student's answers to questions are mostly correct.

2 points Student's representation of the steps in the water treatment process is somewhat inaccurate or incomplete. Student leaves out 2 to 3 steps. Student labels some of the steps accurately. Student's answers to the questions are mostly inaccurate or incomplete.

1 point Student identifies and labels some but not all of the steps in the water treatment process, leaving out 4 or more steps. Student does not answer the questions.

The Water We Use

Make a Diagram

Water used for drinking, cooking, and bathing is made safe by water treatment plants. Make a diagram to show what happens from the time the water enters the treatment plant until it reaches kitchen faucets. Label and describe each step in the process.

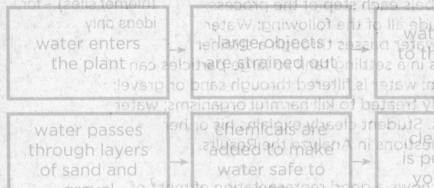
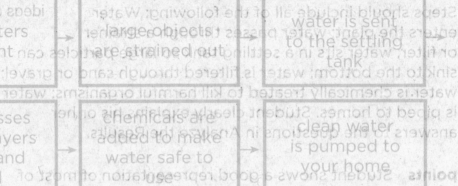

water enters the plant	large objects are strained out	water is sent to the settling tank
water passes through layers of sand and gravel	chemicals are added to make water safe to use	clean water is pumped to your home

Analyze the Results

1. If someone poured oil onto an empty lot down the street, how could this affect your drinking water?

   Chemicals can seep into the groundwater or be washed in

   runoff from heavy rains into streams. These chemicals could

   get into our drinking water and into the water supply and

   harm fish, animals, and plants.

2. Is it more difficult to filter liquid or solid substances out of the water in a treatment plant? Explain.

   It would be more difficult to filter liquid substances because

   solids can be caught in the straining process.

Changes in Weather

Write the word or words that best complete each sentence in the spaces below. Words may be used only once.

atmosphere	hurricane	sphere	weather
axis	precipitation	tornado	
clouds	season	water cycle	

1. Water that falls to the ground from the atmosphere is called _____ precipitation _____.

2. What the air is like at a certain time and place is the _____ weather _____ .

3. The layers of gases and tiny bits of dust that surround Earth are called the _____ atmosphere _____ .

4. The path water takes between Earth's surface and the atmosphere is called the _____ water cycle _____ .

5. Earth is shaped like a ball called a(n) _____ sphere _____ .

6. A(n) _____ hurricane _____ is a storm with strong winds and heavy rains that forms over the ocean.

7. Cirrus, cumulus, and stratus are all types of _____ clouds _____ .

8. Winter is the coldest _____ season _____ of the year.

9. Earth spins on its _____ axis _____ .

10. A(n) _____ tornado _____ is a powerful storm that forms over land.

Circle the letter of the best answer for each question.

11. All of the following are examples of precipitation except

   A hail.

   B rain.

   C lightning.

   D snow.

12. What is the term used for the weight of air pushing down on Earth?

   A temperature

   B air pressure

   C humidity

   D wind

13. Which is the warmest season of the year?

   A winter

   B spring

   C summer

   D fall

14. What powers the water cycle?

   A the Sun

   B the Moon

   C Earth's rotation

   D wind speed

15. A storm that buries cars and buildings in snow is a

   A blizzard.

   B hurricane.

   C tornado.

   D flood.

© Macmillan/McGraw-Hill

Answer the following questions.

16. **Interpret Data** Study the information in the tables below. What relationship do you see between temperature and snowfall?

Chicago	December	January	February
Average Temperature	35°F	14°F	28°F
Average Snowfall	4 in.	28 in.	16 in.

New York	December	January	February
Average Temperature	42°F	35°F	46°F
Average Snowfall	1 in.	8 in.	0 in.

January shows the lowest average temperatures for both

cities and the largest amounts of snowfall.

17. **Using Numbers** Using the same chart, Chicago was how much colder than New York in February?

The difference between average February temperatures in

Chicago and those in New York is 18 degrees F.

18. **Predict** There are many cirrus clouds in the sky. What do you think will happen to the weather?

Cirrus clouds are usually seen during fair weather, but they

could also mean rain in a day or two.

Name _____   Date _____

Answer the following questions.

19. Why does the air temperature usually change throughout the day?

The Sun's energy heats Earth's land and water. In turn, the

land and water heat the air. The Sun is hottest at midday,

which is why the temperature varies throughout the day.

20. What are some similarities between hurricanes and blizzards?

They both contain heavy winds, cause damage to plants,

animals and buildings, and bring lots of precipitation.

21. Critical Thinking Explain how the angle of the Sun's rays affect the climate of a place.

In places where the angle of the Sun is direct, the

temperatures will be hotter because the Sun is striking straight

on. In places where the angle of the Sun is slanted, the

temperatures will be colder because the Sun's rays are not direct.

22. Thinking Like a Scientist A weather report states that a high pressure system is moving into the area. Would this be a good day to plan a picnic? Why or why not?

Yes, it would be a good day to plan a picnic because a high

pressure system means that fair weather will be moving into

the area.

Circle the letter of the best answer for each question.

1. What is the measure of how hot or how cold something is?

   A boiling point

   B air pressure

   C thermometer

   D temperature

2. What is the freezing point of water?

   A 12°F

   B 10°F

   C 32°F

   D 0°F

3. What is a tool for measuring wind speed called?

   A barometer

   B weather vane

   C thermometer

   D anemometer

4. Which of the following is <u>not</u> a form of precipitation?

   A snow

   B sleet

   C hail

   D wind

Critical Thinking Study the chart below. Which city might be best for building a snowman? Why?

City	Average Temperature in January	Average Yearly Precipitation
Albany, NY	22.2°F	38.6 inches
Austin, TX	50.2°F	33.65 inches
Denver, CO	26.1°F	8.99 inches
Reno, NV	33.6°F	7.48 inches

Albany might be the best because its average temperature

is below freezing in January and gets the most precipitation.

Denver's average is also below freezing, but it gets less

precipitation.

Name _____ Date _____

Circle the letter of the best answer for each question.

1. Stratus clouds that form near the ground are called

   A steam.

   B water vapor.

   C fog.

   D cumulus clouds.

2. What is water vapor?

   A water in its gaseous form

   B heated water

   C water in its liquid form

   D cooled water

3. What is condensation?

   A a type of cloud

   B changing of a gas to a liquid

   C the path water takes

   D changing of a liquid to a gas

4. What is another name for white, puffy clouds with flat bottoms?

   A stratus clouds

   B fog

   C cirrus clouds

   D cumulus clouds

Critical Thinking What role does condensation play in the water cycle, and how is this role important?

Condensation causes clouds to form when water droplets collect

on dust. Without clouds, rain would not fall and Earth would run

out of water, ending the water cycle.

_____

_____

_____

Circle the letter of the best answer for each question.

1. The pattern of weather at a certain place over a period of time is called its

    A temperature.

    B climate.

    C precipitation.

    D seasons.

2. What happens to places on Earth where the Sun's rays strike at an angle?

    A They are rainy.

    B They are dry.

    C They are warmer.

    D They are cooler.

3. Which describes climate?

    A temperature and precipitation

    B precipitation and seasons

    C seasons and temperature

    D snow and sun

4. Which city is most likely to be the coldest?

City	Elevation
Austin, TX	501 feet
Billings, MT	3,124 feet
Cheyenne, WY	6,067 feet
New Orleans, LA	11 feet

    A Austin, TX

    B Billings, MT

    C Cheyenne, WY

    D New Orleans, LA

Critical Thinking Imagine you are traveling east from the Pacific Ocean and you have just crossed over a mountain range. What do you think the climate will be like on the eastern side of the mountains? Why?

The climate will be dry, because the mountains will block the

moist ocean air and rain clouds.

Changes in Weather

Write the word or words that best complete each sentence in the spaces below. Words may be used only once.

atmosphere	hurricane	sphere	weather
blizzard	precipitation	tornado	
cloud	seasons	water cycle	

1. A(n) _____ is a collection of tiny water drops or ice crystals in the air.

2. Snow, sleet, and rain are examples of _____ .

3. A(n) _____ has a ball-like shape.

4. The air temperature is one characteristic of _____ .

5. Water moves from Earth's surface into the atmosphere and back again in a process called the _____ .

6. Times of the year with different weather patterns are _____ .

7. A(n) _____ is a long, severe snowstorm.

8. The air that surrounds Earth is part of the _____ .

9. A funnel cloud is a distinct characteristic of a(n) _____ .

10. When a(n) _____ moves from the ocean onto land, its strong winds and heavy rains can cause severe damage.

Circle the letter of the best answer for each question.

11. Where do hurricanes form?

    A over land

    B over mountains

    C over oceans

    D over lakes

12. Which of these tools measures air pressure?

    A an anemometer

    B a rain guage

    C a weather vane

    D a barometer

13. What is precipitation?

    A a collection of tiny water drops seen in the air

    B water that falls to the ground from the atmosphere

    C layers of gases and dust that surround Earth

    D the measurement of how hot or cold something is

14. Which of the following affects climate?

    A sand and rocks

    B windsocks and air

    C weather vanes and balloons

    D oceans and mountains

15. A powerful storm that forms over land and looks like a funnel is a

    A hurricane.

    B twister.

    C tornado.

    D blizzard.

Answer the following questions.

16. **Using Numbers** At what temperature does water freeze?

    A 50°F

    C 32°F

    B 0°F

    D 10°F

17. **Predict** It is raining outside and the temperature has suddenly dropped to 25°F. What do you think will happen and why?

    The rain will probably change to snow. Once the temperature

    drops below the freezing level of 32°F, the precipitation will

    turn to snow.

18. **Interpret Data** Study the information on the tables below. Which city received the least amount of rainfall and when? Which city received the most amount of rainfall and when?

Seattle	March	April	May
Average Temperature	41°F	49°F	53°F
Average Rainfall	4 in.	28 in.	16 in.

New York	March	April	May
Average Temperature	42°F	53°F	67°F
Average Rainfall	17 in.	15 in.	3 in.

New York received the least amount of rainfall in May. Seattle

received the most amount of rainfall in April.

Answer the following questions.

19. The temperature outside is 80°F at 6:00 P.M.
Will the temperature rise or drop by 10:00 P.M.?

The temperature will drop by 10:00 P.M. because by that time

in the evening the Sun will have set. Without the Sun's energy,

the air temperature cools.

20. Explain what a tornado is.

A tornado is a small, powerful storm that forms over land. It

looks like a big, tall funnel. Tornados can be very destructive.

Tornados have powerful winds that move in a circle.

21. Critical Thinking The equator is the imaginary
circle around the middle of Earth. The North Pole
and South Pole are located closest to Earth's axis.
Would a location near the equator receive warmer
weather or colder weather than a location near the
North or South Pole? Explain your answer.

A location near the equator would be warmer than one near

the North or South Poles. The Sun's rays strike directly on

locations near or on the equator.

22. Thinking Like a Scientist There is a strong wind
blowing in from the north. You are riding your
bicycle. Would it be easier for you to ride your
bicycle toward the north or toward the south? Why?

It would be easier to ride a bicycle toward the south because

you would be riding with the wind as it works with you.

Name _____ Date _____

Weather Report

Write a Weather Report

Objective: Students will write a weather report of a typical week in a season, describing a variety of weather patterns using terms learned in Chapter 7.

Scoring Rubric

Materials

- poster board
- markers
- colored pencils
- reference materials

4 points   Student selects a season and a week and successfully writes a weather report using at least five of the terms listed. Student clearly and correctly labels all of the terms on his or her weather map. Student clearly explains his or her answers to the questions in Analyze the Results.

3 points   Student selects a season and a week and writes a weather report using some of the terms listed. Student labels most of the terms correctly on his or her weather map. Student answers the questions in Analyze the Results. Answers to the questions are partially correct.

2 points   Student attempts to write a weather report using some of the terms listed. Student does not label the terms correctly on his or her weather map. Student attempts to answer the questions in Analyze the Results. Answers to the questions are incorrect.

1 point   Student attempts to write a weather report but is unable to use the terms correctly. Student is unable to correctly label the terms on his or her weather map. Student does not answer the questions.

Weather Report

Communicate

Write a weather report describing a variety of weather patterns that may occur in your town during a typical week in a season you choose. Be sure to use at least five of the terms listed below in your report. Use a map of your state to mark and label specific weather patterns that affect your town and where these patterns may occur.

air pressure	hurricane	tornado
cirrus clouds	precipitation	weather
cumulonimbus clouds	stratus clouds	wind
cumulus clouds	temperature	

Analyze the Results

1. Does your map show a high-pressure system or a low-pressure system? How can you tell?

   If a student's map shows a high-pressure system, his or her

   map and report will show fair weather. If a student's map

   shows a low-pressure system, his or her map and report will

   show inclement weather.

2. Choose one specific weather event that may occur in your area and describe it. How would you prepare for it?

   Answers will vary depending on the weather event selected.

Planets, Moons, and Stars

Write the word or words that best complete each sentence in the spaces below. Words may be used only once.

axis	phases	solar system	telescope
constellation	planets	space probe	
orbit	rotates	star	

1. A star and all the objects that move around it is

   a(n) __solar system__ .

2. It takes Earth about 365 days to __orbit__ the Sun.

3. We can see stars by using a(n) __telescope__ .

4. Earth spins on a tilted __axis__ .

5. The different shapes of the Moon that we see are

   called its __phases__ .

6. Day and night happen because Earth __rotates__ .

7. A(n) __space probe__ can explore the Moon and send pictures back to Earth.

8. Mercury, Venus, Earth, and Mars are all inner __planets__ .

9. The Sun is a medium-sized __star__ that provides heat for Earth.

10. Orion is a(n) __constellation__ that we can see in winter.

Circle the letter of the best answer for each question.

11. At what time of day is the Sun highest in the sky?

   A 8 A.M.

   B 12 P.M.

   C 4 P.M.

   D 8 P.M.

12. Which phase of the Moon follows the full moon?

   A dark moon

   B first-quarter moon

   C last-quarter moon

   D new moon

13. We have seasons because of

   A Earth's tilt and revolution around the Sun.

   B Earth's tilt and rotation on its axis.

   C Earth's rotation and the Sun's revolution.

   D Earth's rotation and the Sun's rotation.

14. In June, the Northern Hemisphere begins

   A winter.

   B spring.

   C summer.

   D fall.

15. Each lunar cycle lasts for about

   A 7 days.

   B 14 days.

   C 29 days.

   D 365 days.

Answer the following questions.

16. **Using Numbers** About how many lunar cycles occur in one year? In two years?

There are approximately 12 lunar cycles a year; 24 in two years.

17. **Interpret Data** Study the information in the table below. What relationship do you see between a planet's speed of rotation and its length of day?

Planets and their Rotations

	Diameter of Planet	Speed of Rotation	Length of the Day
Planet A	3,000 miles	29 miles/second	24 hours
Planet B	2,990 miles	60 miles/second	14 hours
Planet C	3,020 miles	120 miles/second	8 hours

The faster a planet spins and orbits the Sun, the shorter the

length of its day.

18. **Infer** Why does Mercury have a shorter year than Neptune?

Mercury is the closest planet to the Sun and has the shortest

orbit. Neptune is the farthest planet away from the Sun. It

takes much longer for Neptune to orbit the Sun.

Answer the following questions.

19. Why are shadows the shortest at noon?

At noon, the Sun is directly overhead and shines light at an

angle that creates the shortest shadow.

20. Would it be easier to see stars in the night sky in a city or in the country? Explain your answer.

It would be easier to see stars in the country at night because

there are not as many lights from buildings and streetlights

to light up the night sky. The country sky appears darker and

allows more stars to be seen.

21. Critical Thinking Is it true that only one side of the Moon receives sunlight, while the other always remains dark? Explain your answer.

One side of the Moon always receives light at any given time.

Since the Moon rotates, all sides of the Moon will eventually

receive sunlight as it orbits Earth.

22. Thinking Like a Scientist You hypothesize that constellations appear to move in the night sky because of Earth's orbit. Describe what you would use and how you might test your hypothesis.

Answers may include: I would use a telescope to view the Big

Dipper and Little Dipper. I would make a chart to show the

time and position in the sky and test it by plotting this data

over several weeks. Then I would try to draw a conclusion.

Circle the letter of the best answer for each question.

1. The Sun is about how many kilometers (km) from Earth?

   A 150 thousand km

   B 100 thousand km

   (C) 150 million km

   D 100 million km

2. What do shadows look like at midday?

   (A) short

   B narrow

   C wide

   D long

3. How long does it take for Earth to make one complete turn on its axis?

   A 12 hours

   (B) 24 hours

   C 30 days

   D 365 days

4. How long does it take for Earth to make a complete orbit around the Sun?

   A 24 hours

   B 7 days

   C 30 days

   (D) 365 days

Critical Thinking What would happen if the Sun and Earth were closer together?

Earth would be very hot and living things on Earth might die.

There is life on Earth because it is the appropriate distance from

the Sun to maintain the right temperature to sustain life.

Circle the letter of the best answer for each question.

1. The shape of the Moon that people see is called

   A a lunar night.

   B a month.

   C a phase.

   D a crescent.

2. What type of Moon is represented below?

   A a new Moon

   B a full Moon

   C a crescent Moon

   D a last quarter Moon

3. What is a satellite?

   A a cycle of the Moon where the whole Moon can be seen

   B anything that revolves around another larger object in space

   C one complete orbit of Earth around the Sun

   D a vehicle that transports astronauts into space

4. Why does the Moon appear to shine at night?

   A it makes its own light

   B sunlight bounces off it

   C it has a glowing ring around it

   D it reflects light from Earth

Critical Thinking How can the phases of the Moon be compared to a calendar?

Each cycle of the Moon repeats in 29 days, which is about one

calendar month. Each phase of the Moon repeats in order during

each cycle. A person can rely on the Moon much like he or she

relies on a calendar.

Circle the letter of the best answer for each question.

1. What is the center of our solar system?

   A the Sun
   B the Moon
   C Earth
   D Saturn

2. What tool can people use to view planets in faraway space?

   A microscope
   B telescope
   C periscope
   D satellite

3. What is the largest planet?

   A Mercury
   B Earth
   C Saturn
   D Jupiter

4. Study the diagram below.

   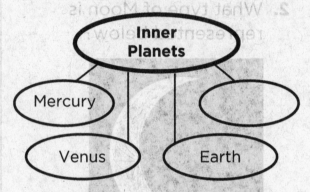

   Which of these belongs in the empty circle?

   A Uranus   **C** Mars
   B Jupiter   **D** Neptune

Critical Thinking It takes Earth about 365 days to complete an orbit. Would it take Saturn more time or less time to complete its orbit? Explain your answer.

It would take longer because the farther away a planet is from the

Sun, the longer it takes to complete its orbit.

_____

_____

Circle the letter of the best answer for each question.

1. What is a hot, glowing ball of gas that gives off light and heat?

    A a planet

    B a meteor

    C a star

    D a moon

2. How many stars are in our solar system?

    A one

    B five

    C thousands

    D millions

3. In what season can the constellation Orion be seen?

    A spring

    B summer

    C fall

    D winter

4. What constellation is represented in the picture?

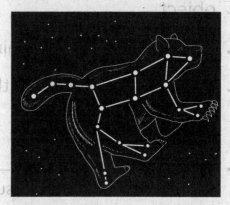

    A Orion

    B Little Dipper

    C Ursa Major

    D the hunter

Critical Thinking If you were stargazing and observed a blue star, what would you know about it?

If I saw a blue star, I would know that it was a very hot star since

blue stars are the hottest of all stars.

© Macmillan/McGraw-Hill

Planets, Moons, and Stars

Write the word or words that best complete each sentence in the spaces below. Words may be used only once.

axis	orbit	revolve	star
constellation	phases	rotates	
craters	planet	solar system	

1. A hot, glowing ball of gases is a(n) _____ .

2. To _____ means to move around another object.

3. The Moon has eight main _____ .

4. A pattern or picture outlined by stars is called a(n) _____ .

5. The Moon has many _____ formed by rocks crashing into its surface.

6. A(n) _____ is the regular path the Earth travels on around the Sun.

7. The Earth _____ or spins in space.

8. A real or imaginary line through the center of a spinning object is called its _____ .

9. The planets are part of our _____ .

10. A(n) _____ is a large body of rock or gas that revolves around a star.

© Macmillan/McGraw-Hill

Circle the letter of the best answer for each question.

11. The Sun appears to travel in the sky from

A west to east.

B east to west.

C south to north.

D north to south.

12. In December, the Northern Hemisphere

A is tilted away from the Sun.

B is tilted towards the Sun.

C gets the most heat from the Sun.

D gets the most light from the Sun.

13. The Moon appears to glow in the sky because

A it generates its own light.

B it reflects light from the Sun.

C it reflects light from Earth.

D it absorbs light from constellations.

14. In which of the following months are the days shortest in the Northern Hemisphere?

A December

B March

C September

D June

15. Which phase of the Moon comes directly before a new Moon?

A full moon

B crescent moon

C last-quarter moon

D gibbous moon

Answer the following questions.

16. Interpret Data Study the information in the table below. Which planet has the shortest day?

Length of Day

Planet	Day = time for complete spin (in Earth hours or days)
Mercury	59 days
Venus	243 hours
Earth	24 hours
Mars	24 hours, 37 minutes
Jupiter	9 hours, 56 minutes
Saturn	10 hours, 49 minutes
Uranus	17 hours, 14 minutes
Neptune	16 hours, 7 minutes

Jupiter has the shortest day with 9 hours, 56 minutes.

17. Infer Does Earth have a longer or shorter year than Mercury?

Earth is farther away from the Sun than Mercury. It takes

longer for Earth to orbit the Sun than it does for Mercury.

18. Using Numbers How many phases of the Moon occur in each lunar cycle? How many in two months?

There are 8 main phases in each lunar cycle, which occurs

once a month; about 16 phases in two months.

© Macmillan/McGraw-Hill

Answer the following questions.

19. Although stars are always present in the sky, why is it that people cannot see them during the day?

People cannot see stars during the day because the Sun's

bright light blocks the stars' light from our view.

20. What happens to shadows at sunrise and sunset? How are these shadows different from those cast at noon?

Both at sunrise and at sunset, the Sun is low in the sky and

its angle casts long shadows. At noon, the Sun is directly

overhead, so shadows are short.

21. Critical Thinking Explain how the Moon seems to change shape throughout its cycle.

As the Moon revolves around Earth, different amounts of the

Moon's lighted half face Earth. The lighted sections that can

be seen from Earth are the phases that appear like changing

shapes.

22. Thinking Like a Scientist What might be a good way to find a particular star in the night sky?

Answers will vary but may include: Use a telescope and find a

constellation like Orion. Once it has been located, a particular

star can be traced in relation to where it is in the sky to the

constellation.

Solar System Poster

Make a Solar System Poster

Objective: Students will make and label a poster of the solar system, showing the planets and their correct positions around the Sun.

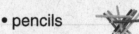
Scoring Rubric

4 points Student completes the activity, accurately draws and labels each planet in its proper position around the Sun. Student clearly explains his or her answers to the questions in Analyze the Results.

3 points Student completes the activity, draws each planet, but may have mislabeled one or two planets or placed one or two planets incorrectly around the Sun. Student answers the questions in Analyze the Results. Answers to the questions are mostly correct.

2 points Student completes the activity, draws each planet, but may have mislabeled three or four planets or placed three or four planets incorrectly around the Sun. Student attempts to answer the questions in Analyze the Results. Answers to the questions are mostly incorrect.

1 point Student completes part of the activity, but his or her drawings and placements of planets are mostly incorrect. Student does not answer the questions or most of the answers are incomplete or incorrect.

Solar System Poster

Communicate

Use the materials your teacher has given you to make
a poster of our solar system. Design the poster to
show the names of the planets. Place them in their
correct position around the Sun. Be sure to show the
path of the planets in their orbits around the Sun.

Analyze the Results

1. What planet is farthest from the Sun? Why might
this change? Explain.

Neptune is the farthest from the Sun. Pluto is no longer

considered a planet and scientists have seen objects like

UB313 that may one day be considered as another planet.

2. How might a planet's distance from the Sun affect
conditions on that planet?

Answers will vary but may include: The farther away a planet

is from the Sun, the colder it will be; it may be too cold to

have water or living things. If it is close to the Sun, it may be

too hot to have water or support living things.

3. What other objects can be found in our solar
system? Where would you place them on your
poster?

Answers will vary. Accept all reasonable answers.

Name _____   Date _____

Observing Matter

Write the word or words that best complete each sentence in the spaces below. Words may be used only once.

element	liquid	metric system	weight
gas	mass	solid	
gravity	matter	volume	

1. Matter that has a definite size and shape is a(n) _____ .

2. Matter that is not solid or gas is a(n) _____ .

3. The pull of gravity on an object is called _____ .

4. Matter that has no definite shape or volume is a(n)

   _____ .

5. Anything that takes up space is _____ .

6. Scientists use the _____ , which is a
   system of weights and measures.

7. A "building block" that makes up all matter is

   called a(n) _____ .

8. The amount of space an object takes up is called

   its _____ .

9. A bowling ball is heavier than a beach ball because

   it has more _____ .

10. The force that keeps objects from floating into

    space is called _____ .

Circle the letter of the best answer for each question.

11. Which of the following <u>best</u> describes matter?

  A anything that has mass and volume

  B anything that is solid

  C anything that can be seen

  D anything that is in the air

12. Which of the following statements is false?

  A Mass is a measure of the amount of matter in an object.

  B Objects with the same volume always have the same mass.

  C In the metric system, mass is measured in grams.

  D A marble has more mass than a piece of popcorn.

13. Which is <u>not</u> a state of matter?

  A gravity

  B gas

  C solid

  D liquid

14. Which state of matter has the most energy?

  A solid

  B liquid

  C gas

  D weight

15. What is one special property of a magnet?

  A it has mass

  B it is made of wood

  C it can float

  D it can attract certain metals

Answer the following questions.

Study the table shown below for measuring the volume of solid objects.

Object	Water Level Before	Water Level After
rock	10 liters	17 liters
marble	12 liters	14 liters
coins	11 liters	16 liters

16. **Interpret Data** What is the volume of the marble? How do you know?

The marble's volume is 2 liters. I know this because the water level increased by 2 liters, which means the marble has a volume of 2 liters.

17. **Infer** Which object has the greatest volume? Why?

The rock has the greatest volume because it displaced the most water.

18. **Predict** What would happen if you added another rock of the exact same size to the water?

The water level would increase by another 7 liters bringing the total water level to 24 liters.

Answer the following questions.

19. Why would a person weigh less on the Moon than he or she does on Earth?

The pull of gravity between the person and the Moon is

weaker than the pull between the person and Earth.

20. Why is a wooden spoon better for cooking than a metal spoon?

Wood does not heat up as quickly as metal. Therefore, a

wood spoon would stay cooler and be easier to handle when

cooking.

21. **Critical Thinking** Explain why some objects float in water while others sink. Give an example of an object that sinks and one that floats.

Answers will vary depending on objects chosen but should

explain that objects with little mass and a lot of volume are

more likely to float than objects with a lot of mass and little

volume.

22. **Thinking Like a Scientist** What tool should a student use to measure the mass of a golf ball? Why?

A student should use a pan balance to measure the mass of a

golf ball. A golf ball's mass is easily measured by placing another

object whose mass is known on the other side of the balance.

Circle the letter of the best answer for each question.

1. The way an object looks, tastes, smells, sounds, and feels are called

   A volumes.

   B measures.

   C properties.

   D observations.

2. Why do objects float?

   A because they have a lot of mass and little volume

   B because they have equal amounts of mass and volume

   C because they have little mass and a lot of volume

   D because they have all mass and no volume

3. Which of the following would be the best object for stirring hot liquids?

   A a wooden spoon

   B an iron spoon

   C a copper spoon

   D a steel spoon

4. How many different elements are there that make up all of the matter in the world?

   A less than 10

   B more than 100

   C more than 1,000

   D more than 10,000

Critical Thinking If two different matters are made of the same elements, how is it possible that they look and feel different?

They may look and feel different because they may be made up of

different combinations of the same elements in different amounts.

_____

_____

_____

Use with **Lesson 1**
Properties of Matter

Name _____ Date _____

Circle the letter of the best answer for each question.

1. What unit of measure do people agree to use?

    A metric system

    B Fahrenheit

    C Celsius

    D standard unit

2. In the metric system, what are units of length based on?

    A meter

    B liter

    C gram

    D inch

3. Which instrument will best measure mass?

    A graduated cylinder

    B thermometer

    C pan balance

    D tape measure

4. In the metric system, what unit is used to measure a liquid's volume?

    A meter

    B liter

    C gram

    D kilo

Critical Thinking

Which of the three units of measure— meter, centimeter, or millimeter—would be best for measuring the football field? Why?

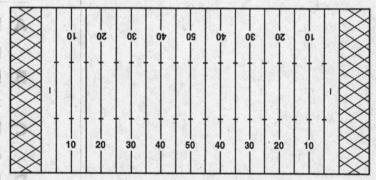

The meter would be best for measuring the football field because

it is the longest unit and would provide the most accurate

measurement.

© Macmillan/McGraw-Hill

Name _____   Date _____

Circle the letter of the best answer for each question.

1. What is the scientific term for the forms of matter?

   A units of matter

   B types of matter

   C states of matter

   D properties of matter

2. How many forms of matter exist on Earth?

   A 30

   B 3

   C 100

   D too many to count

3. Look at the illustration below. What form of matter does illustration B represent?

   A   B   C

   A cube

   B solid

   C liquid

   D gas

4. All of the following are characteristics of liquids except

   A liquid particles can slide past each other.

   B liquids cannot change shape.

   C liquids take up a definite amount of space.

   D liquids take up the shape of their containers.

Critical Thinking Describe how you use the three states of matter in your life every day.

Answers will vary but should include: using solid form—food;

liquid form—water; and gas form—the air we breathe.

Observing Matter

Write the word or words that best complete each sentence in the spaces below. Words may be used only once.

element	liquid	metric system	weight
gas	mass	pan balance	
gravity	matter	solid	

1. A(n) _____ is a form of matter that has no definite shape or volume.

2. A tool called a _____ measures mass.

3. Milk can be defined as a(n) _____ .

4. When scientists measure, they use the _____ .

5. Matter that has shape and hardness is a(n) _____ .

6. The force of _____ keeps objects from floating in space.

7. The amount of gravity that is needed to keep an object on Earth is its _____ .

8. A balloon has less _____ than a basketball.

9. A material from which all other materials are made is a(n) _____ .

10. All things are made up of _____ .

Circle the letter of the best answer for each question.

11. Which of the following will a magnet attract?

   A glass

   B plastic

   C certain metals

   D wood

12. All matter is made up of

   A liquids

   B solids

   C elements

   D gases

13. Which is the <u>best</u> definition of volume?

   A how much mass an object has

   B how much space an object takes up

   C how much matter in an object

   D how much metal in an object

14. Which of the following statements is true?

   A A bowling ball has more mass than a tennis ball.

   B In the metric system, mass is measured in inches.

   C Mass is a measure of the amount of liquid in an object.

   D Objects with the same volume always have the same mass.

15. Water, milk, and juice are examples of

   A weights.

   B solids.

   C liquids.

   D gases.

Answer the following questions.

Study the chart shown below.

Volume of Solid Objects

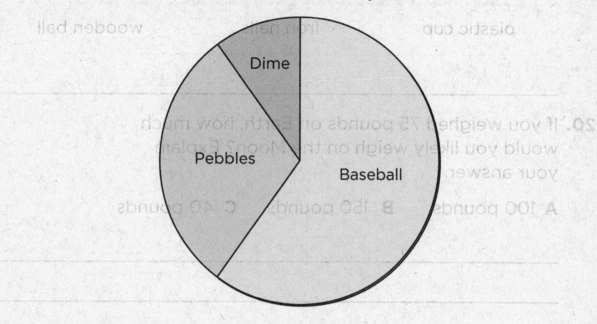

16. Infer Which object has the least volume?

the dime

17. Predict If you were to measure a nickel, do you think it would have more volume or less volume than a dime? Why?

A nickel would have more volume because it is bigger in size

and has more weight than a dime.

18. Interpret Data How would you rank the items from least to greatest volume if you had four dimes instead of one dime?

pebble, dimes, and baseball

Answer the following questions.

19. Of the following objects, which would best conduct heat? Why?

| plastic cup | iron nails | wooden ball |

The iron nails. Heat moves easily through metals such as iron.

20. If you weighed 75 pounds on Earth, how much would you likely weigh on the Moon? Explain your answer.

A 100 pounds   **B** 150 pounds   **C** 40 pounds

C. 40 pounds. The pull of gravity between you and the Moon

is weaker than the pull between you and Earth, so you would

weigh less on the Moon than you do on Earth.

21. **Critical Thinking** Of the list of objects below, circle the ones that are more likely to sink in water.

balloon   gold ring   paper cup   stone
baseball   leaf   beach ball
feather   marble   quarter

22. **Thinking Like a Scientist** Why should scientists use the same units of measure for their experiments?

All scientists should use the same standard units of measure

for their experiments so that all measurements will be

consistent with one another.

© Macmillan/McGraw-Hill

States of Matter

Objective: Student will group objects according to their state of matter: solid, liquid, or gas.

Scoring Rubric

4 points Student shows a good representation of grouping all objects in all three categories. Student clearly identifies properties of all of the objects. Student clearly explains his or her answers to the questions in Analyze the Results, and answers are correct.

3 points Student shows a good representation of grouping most of the objects in all three categories. Student clearly identifies properties of most of the objects. Student's answers to the questions in Analyze the Results are mostly correct.

2 points Student's representation of the grouping of objects is somewhat inaccurate. Student identifies properties of some of the objects accurately. Student's answers to the questions in Analyze the Results are mostly inaccurate or incomplete.

1 point Student identifies and groups some but not all of the objects. Student does not answer the questions in Analyze the Results.

Materials

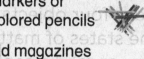

- poster board
- markers or colored pencils
- old magazines and newspapers (for cutting out pictures)
- glue
- scissors
- reference materials (such as encyclopedias)

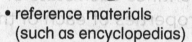

States of Matter

Classify

Choose four objects that represent examples of each of the states of matter (solid, liquid, gas). Use the library or media center to find information about the objects. Draw or cut out pictures from magazines or newspapers of the objects that are to be represented. Then make a chart to classify each object under its proper state. Be sure that your chart correctly identifies each state of matter. Be sure to note the properties of each of the objects you classified.

Analyze the Results

1. Why did you choose to classify the objects the way that you did?

   Answers will vary depending on the objects selected.

   _____

   _____

   _____

2. Are there any objects that could fit into more than one category? Explain your answer.

   Answers will vary depending on the objects selected.

   _____

   _____

   _____

   _____

© Macmillan/McGraw-Hill

Changes in Matter

Write the word or words that best complete each sentence in the spaces below. Words may be used only once.

boils	freeze	solution
chemical change	melts	water vapor
condenses	mixture	
evaporates	physical change	

1. The gaseous state of water is called _____.

2. Water will _____ when it gets cold enough—changing it from a liquid to a solid.

3. When one or more types of matter are mixed evenly with another type of matter, a(n) _____ forms.

4. When a solid changes into a liquid, it _____.

5. When a gas cools and becomes a liquid, it _____.

6. Matter looks different after a(n) _____, but it is still made up of the same kind of matter.

7. When a material changes into a new kind of matter, a(n) _____ takes place.

8. When water is heated and bubbles form, it _____.

9. When you make fruit salad, you create a(n) _____.

10. When a liquid changes into a gas without boiling, it _____.

Name _____ Date _____

Circle the letter of the best answer for each question.

11. What kind of change takes place when sand is made into a sandcastle?

   A a change in solution

   B a physical change

   C a chemical change

   D a change in temperature

12. All of the following represent types of physical changes <u>except</u>

   A change in color.

   B change in shape.

   C change in state.

   D change in size.

13. All of the following represent types of chemical changes <u>except</u>

   A change in temperature.

   B change in color.

   C change in shape.

   D the appearance of bubbles.

14. What does all matter have in common?

   A All matter is high in energy.

   B All matter is made up of tiny particles.

   C All matter is held tightly together.

   D All matter is solid.

15. What happens to water when it freezes?

   A It turns into water vapor.

   B It turns into gas.

   C It takes up less space.

   D It takes up more space.

Name _____ Date _____

Answer the following questions.

Use the diagram below to answer the following question.

16. **Interpret Data** Look at the pictures below. For
    each picture, tell whether a physical change or a
    chemical change has taken place.

chemical change

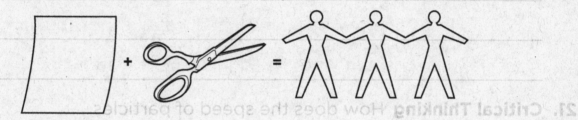

physical change

17. **Make a Model** Draw and label examples of the
    three forms of water.

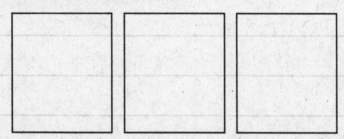

Drawings should include
water (labeled "water"),
ice (labeled "ice"), and
steam or vapor (labeled
"water vapor" or "gas").

18. **Infer** Is an apple turning brown an example of a
    physical change or a chemical change? Explain.

Chemical change; a change in color represents a chemical

change.

© Macmillan/McGraw-Hill

Answer the following questions.

19. What is the difference between a physical change and a chemical change?

When a substance undergoes a physical change, its chemical properties and makeup do not change. In a chemical change, it becomes a new type of matter; its properties and makeup change.

20. Why does dew form on a plant's leaves on a cool morning?

The water vapor in the air touches a cool object (such as a leaf) and loses energy. This causes the water vapor to change its state, thus changing into its liquid form of water.

21. Critical Thinking How does the speed of particles in matter affect the state of the object?

When the particles have the least amount of energy, a solid is formed, such as in ice. In a liquid state they have more energy, allowing the particles to flow around one another. In a gas state they have the most energy, allowing the particles to move freely.

22. Think Like a Scientist If a baker is making bread and wants the yeast in the dough to rise, should he put the dough in the refrigerator or in a warm room? Why?

In a warm room, the rise in temperature will stimulate the particles to move, allowing the dough to rise.

© Macmillan/McGraw-Hill

Circle the letter of the best answer for each question.

1. What happens when matter is heated?

   (A) it gains energy

   B all energy is removed

   C it loses energy

   D energy is split

2. What happens when gas condenses?

   A it becomes hot

   (B) it becomes a liquid

   C it becomes frozen

   D it becomes a solid

3. All of the following are states of water <u>except</u>

   A ice

   B liquid water

   (C) dry ice

   D water vapor

4. Which thermometer shows water in its solid state?

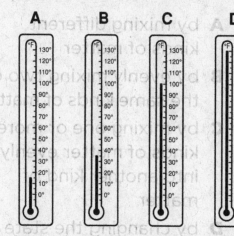

   (A) thermometer A

   B thermometer B

   C thermometer C

   D thermometer D

Critical Thinking Which would take up more space in a container—16 ounces of liquid water or 16 ounces of frozen water? Why?

16 ounces of frozen water because it expands and takes up more

space than liquid water

Circle the letter of the best answer for each question.

1. All of the following represent physical changes in paper <u>except</u>

   A burning.

   B cutting.

   C folding.

   D crumbling.

2. How is a solution formed?

   A by mixing different kinds of matter

   B by evenly mixing two of the same kinds of matter

   C by mixing one or more kinds of matter evenly into another kind of matter

   D by changing the state of two or more kinds of matter

3. All of the following are ways mixtures can be separated <u>except</u>

   A by evaporation

   B by its properties

   C by filters

   D by weight

4. Study the table below.

Mixtures
vegetable soup
salad dressing
clouds

   What belongs in the empty box?

   A brass

   B salt water

   C chocolate milk

   D salad

Critical Thinking How can magnets be used to separate mixtures?

A magnet will attract any object that is magnetic. Therefore, any

objects with magnetic properties can be separated and collected

by the magnet.

Circle the letter of the best answer for each question.

1. What is one way you can tell that a chemical change has taken place?

   A the object changes size

   B the object changes color

   C the object changes weight

   D the object changes shape

2. How does your body go through chemical changes?

   A by growing

   B by sleeping

   C by breaking down food

   D by learning new skills

3. What has to occur for cake batter to undergo a chemical change?

   A it has to be heated

   B it has to be cooled

   C it has to be mixed

   D it has to be frozen

4. Which of the following represents a chemical change of an apple?

   A peeling it

   B cooking it

   C turning brown

   D falling off a tree

Critical Thinking How is rust the result of a chemical change?

Iron becomes rust as a result of being exposed to elements like

oxygen and water. The iron then becomes weaker and changes

color, becoming rust.

_____

_____

Changes in Matter

Write the word or words that best complete each sentence in the spaces below. Words may be used only once.

boil	freeze	solution
chemical change	melt	water vapor
condenses	mixture	
evaporates	physical change	

1. When water _____ on cool mornings it is called dew.

2. Salt water is a(n) _____ because salt and water are mixed evenly.

3. Water in the form of gas is called _____ .

4. A liquid _____ when it changes into a gas without boiling.

5. When a log burns, a(n) _____ takes place.

6. If a liquid is cooled enough, it will _____ or change from a liquid to a solid.

7. Adding milk to cereal creates a(n) _____ .

8. If a rock is exposed to extremely high heat, it will _____ .

9. Tearing a piece of paper is an example of a(n) _____ in the paper.

10. If water is heated to a certain temperature, it will _____ .

Circle the letter of the best answer for each question.

11. Which of the following represents a chemical change?

   A a change in color

   B a change in shape

   C a change in size

   D a change in state

12. What happens to water when it evaporates?

   A it melts

   B it condenses

   C it turns into vapor

   D it freezes

13. Which of the following is not a mixture?

   A salt and pepper

   B milk

   C salad

   D cake

14. Which of the following represents a physical change?

   A copper turns green

   B iron turns to rust

   C dough is baked into bread

   D water freezes into ice

15. What kind of change takes place when a banana turns brown?

   A a chemical change

   B a change in shape

   C a physical change

   D a change in temperature

Answer the following questions.

16. **Make a Model** Draw and label three ways paper can go through physical or chemical changes.

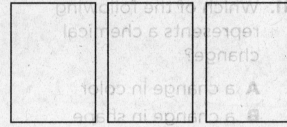

Drawings and labels will vary but may include: cut paper: physical change; torn paper: physical change; burned paper: chemical change

17. **Infer** Is rain that turns to snow an example of a physical change or a chemical change? Explain your answer.

Rain changing to snow represents a physical change in state.

Although snow looks different from rain, it is still made up of

the same kind of matter.

18. **Interpret Data** Label what state water would be in at each of the temperatures shown below.

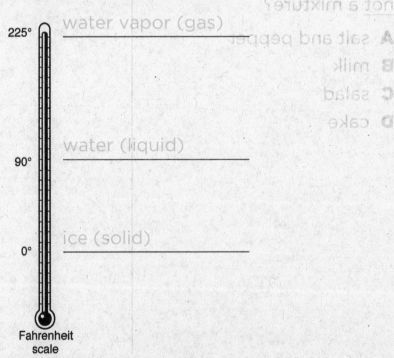

water vapor (gas)

225°

water (liquid)

90°

ice (solid)

0°

Fahrenheit scale

Answer the following questions.

19. What is the scientific difference between cutting a tree limb and burning a tree limb?

Cutting a tree limb is a physical change and burning a tree limb is a chemical change.

20. Describe the similarities and differences between melting and freezing.

Both are physical changes. When a solid gains heat energy, it melts and becomes a liquid. When a liquid loses energy, it freezes and becomes a solid.

21. Critical Thinking How is exercise important in the breaking down of food in your body?

When you exercise, you burn energy. This helps create a chemical reaction that turns the food into fuel for your body.

22. Think Like a Scientist A scientist mixes two different chemicals and wants to know if they will undergo a chemical change or create a mixture. What are the signs the scientist can look for to see what type of change has occurred?

The scientist can look for the formation of gas (shown by the presence of bubbles), the formation of light and heat, or a change in color indicating that a chemical change has occurred. If the chemicals maintain their original properties and can easily be separated, it is a mixture.

Water Brochures

Objective: Student will create a picture book that describes the changes water goes through as heat is added and taken away. Students will include a description of the water as it goes through each state.

Scoring Rubric

4 points Student correctly illustrates and describes all four changes in state using appropriate terms. Student correctly describes the changes as related to adding or taking away heat energy. Student correctly describes the properties of each state. Student clearly and correctly explains his or her answers to the questions in Analyze the Results.

3 points Student correctly illustrates and describes at least three changes in state using appropriate terms. Student's description of the changes as related to adding or taking away heat energy is mostly correct. Student's descriptions of the properties of each state are mostly correct. Student's answers to the questions are mostly correct.

2 points Student illustrates and describes one or two changes in state correctly. Student's description of the change as related to adding or taking away heat is somewhat inaccurate. Student's descriptions of the properties of each state are somewhat inaccurate. Student's answers to the questions are mostly inaccurate or incomplete.

1 point Student's illustrations and description of changes in state are inaccurate. Student's descriptions of the changes as related to adding or taking away heat energy are missing or are mostly inaccurate. Student's descriptions of the properties of each state are missing or mostly inaccurate. Student does not answer the questions.

Water Brochures

Communicate

Design a picture book, showing in both labeled pictures and words, how water changes when heat energy is added and taken away. Be sure to include a description of the water as it goes through each state.

Analyze the Results

1. What state is water in when its temperature is very high? Describe the properties of this state.

   Water is a gas at very high temperatures. As a gas, water has

   no definite volume or shape.

2. Describe the sequence of changes as water goes from a very low temperature to a very high one.

   At cold temperatures, water is a solid. As temperatures rise,

   the water gains energy and the solid melts, becoming a liquid.

   With more energy, the water will evaporate and become a

   gas.

3. Are the changes that water undergoes as it is heated and cooled physical or chemical changes? Explain your answer.

   As water changes states it undergoes physical changes.

   Changing from one state to another is a physical change

   because the makeup of the water has not changed.

Forces and Motion

Write the word or words that best complete each sentence in the spaces below. Words may be used only once.

compound machine	gravity	speed
distance	motion	work
force	position	
friction	simple machine	

1. Any change in position of an object is called
   _____ .
   (motion)

2. An object moves when _____ is
   applied.
   (force)

3. A can opener is an example of a _____ .
   (compound machine)

4. The force that pulls us toward Earth is _____ .
   (gravity)

5. The location of an object is its _____ .
   (position)

6. A device such as a lever is a _____ .
   (simple machine)

7. When a force changes an object's motion,
   _____ is being done.
   (work)

8. The time it takes for something to move from one
   place to another is its _____ .
   (speed)

9. A force that occurs when one object rubs against
   another is _____ .
   (friction)

10. The space between two objects is their _____ .
    (distance)

Circle the letter of the best answer for each question.

11. A magnetic force can attract which of the following?

   A wood

   B glass

   C iron

   D rubber

12. Which of the following is an example of friction?

   A an apple falling from a tree

   B using the brakes on a bicycle

   C using batteries to power a toy

   D water freezing

13. Which of the following is not a simple machine?

   A scissors

   B screw

   C pulley

   D lever

14. A pile of wood has

   A electrical energy.

   B kinetic energy.

   C potential energy.

   D magnetic energy.

15. Which tool measures speed?

   A stopwatch

   B thermometer

   C spring balance

   D ruler

Answer the following questions.

16. **Communicate** Describe your position in the classroom.

Answers will vary but should include words such as: in front of; behind; next to; near, etc.

17. **Interpret Data** Study the table below. Use a combination of foods from the chart to plan a meal that contains between 500 and 650 calories. Show how you arrived at the total.

Food	Calories
1 large slice of bread	100
1 slice of cheese	100
1 apple	50
1 banana	85
chicken	215
mixed salad	150
orange juice	75
milk	115

Accept all reasonable answers that total between 500 and 650 calories.

18. **Infer** Describe why a pair of scissors is a compound machine.

Scissors are made up of two simple machines—levers that move back and forth and wedges that cut the fabric or paper.

Answer the following questions.

19. Choose three of the six simple machines and give examples of how they are used in your daily life.

Answers will vary but should include information on three of

the following machines: lever; pulley; wheel and axle; inclined

plane; wedge; screw.

20. Explain why it would take more force to move a bowling ball than it would to move a beach ball.

It would take more force to move a bowling ball because it is

heavier than a beach ball.

21. Critical Thinking Where are some places stored energy is found?

Answers will vary but may include: food, wood, batteries, a

ball at the top of a hill.

22. Thinking Like a Scientist How can the stored energy from the food we eat be converted into energy of motion?

Stored energy from food is converted into energy of motion

when we burn the calories from the food we eat. Calories

are burned in numerous ways, from running and walking to

brushing our teeth and combing our hair.

Name _____ Date _____

Circle the letter of the best answer for each question.

1. What does the word *position* mean?

  A the speed of an object

  B the location of an object

  C the amount of space between objects

  D how an object moves

2. What two things must be known to measure speed?

  A how far an object traveled and how long it took to go that distance

  B how much an object weighs and how far it traveled

  C how dense an object is and how far it traveled

  D how long an object took to go a certain distance

3. What do the words *over*, *under*, *left*, and *right* give clues to?

  A speed

  B distance

  C position

  D balance

4. At the speed shown, which vehicle would travel 10 miles in the least amount of time?

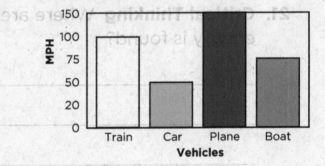

  A train

  B car

  C plane

  D boat

Critical Thinking A student is asked to run 50 yards. He can choose to run it in a straight or zigzag course. Which course would allow him to reach his destination first?

He would reach his destination first by using the straight course

because the fastest distance between two points is a straight line.

Name _____ Date _____

Circle the letter of the best answer for each question.

1. What causes a magnet to attract and repel?

  A magnetic force

  B contact force

  C gravity

  D friction

2. What is gravity?

  A the pulling force between two objects

  B the weight of an object in space

  C the changing state of an object

  D the weight plus the mass of an object

3. All of the following will increase friction <u>except</u>

  A sandpaper.

  B rough stones.

  C oil.

  D rubber.

4. Study the chart. Which object has the greatest weight?

Object	Amount of Gravity
skateboard	7
scooter	15
bicycle	30
canoe	75

  A canoe

  B bicycle

  C scooter

  D skateboard

Critical Thinking Suppose you lost an iron key in the sand. What could you use to help find the key?

I would use a large magnet to help me find the key because the

iron key is magnetic and the magnet would attract the key and

pull it from the sand.

Circle the letter of the best answer for each question.

1. What makes changes in motion possible?

   (A) energy

   B work

   C gravity

   D weight

2. Which activity requires the <u>most</u> work?

Activity	Amount of Energy Used
Jumping rope	moderate
Lifting a book	light
Hammering a nail	heavy
Reading a book	none

   A jumping rope

   B lifting a book

   (C) hammering a nail

   D reading a book

3. What happens when a force changes an object's motion?

   A the object changes state

   (B) work is done

   C gravity is interrupted

   D a machine is being used

4. What are the two main forms of energy?

   A energy of speed and energy of sound

   (B) energy of motion and potential energy

   C energy of force and energy of motion

   D energy of speed and potential energy

Critical Thinking Explain why kicking a soccer ball is work.

Work is done only when a force changes an object's motion. Work is being done when a ball is kicked because the object's motion is changed.

© Macmillan/McGraw-Hill

Circle the letter of the best answer for each question.

1. A doorknob is an example of which simple machine?

    A pulley

    B wedge

    C wheel and axle

    D inclined plane

2. Which simple machine could be used to move a heavy box up some stairs?

    A an inclined plane

    B a lever

    C a wheel and axle

    D a screw

3. How many types of simple machines exist?

    A 4

    B 6

    C 8

    D 12

4. Which simple machine would a person use to lift a heavy rock?

    A wedge

    B lever

    C inclined plane

    D wheel and axle

Critical Thinking Why do public buildings such as schools, libraries, and hospitals have ramps at their entrances?

Possible answer: One reason for having ramps in public buildings

is so that it is easier for wheelchairs to access the buildings.

Ramps are also easier if heavy deliveries need to be made.

Forces and Motion

Write the word or words that best complete each sentence in the spaces below. Words may be used only once.

compound machine	motion	wheel and axle
energy	simple machine	work
kinetic energy	speed	
lever	wedges	

1. Lifting a feather is an example of _____.

2. While an object is changing position it is in _____.

3. All moving objects have _____.

4. Two or more simple machines together make a(n) _____.

5. A door knob is an example of a(n) _____.

6. Most cutting tools are simple machines called _____.

7. A pulley is an example of a(n) _____.

8. The food we eat has stored _____.

9. A straight bar that moves on a fixed point is a(n) _____.

10. How fast an object moves is its _____.

© Macmillan/McGraw-Hill

Circle the letter of the best answer for each question.

11. A pulley is an example of which of the following?

   A axle

   B compound machine

   C simple machine

   D magnet

12. Which word cannot be used to describe an object's position?

   A under

   B tomorrow

   C left

   D beside

13. A magnet can attract all of the following <u>except</u>

   A rubber bands.

   B paper clips.

   C thumb tacks.

   D nails.

14. Which of the following is an example of work?

   A watching television

   B pushing against a concrete wall

   C lifting a pencil

   D studying for a science test

15. Which statement about energy is true?

   A There are three main forms of energy.

   B Stored energy is energy in motion.

   C Energy cannot make matter change.

   D Energy is the ability to do work.

Answer the following questions.

16. **Interpret Data**

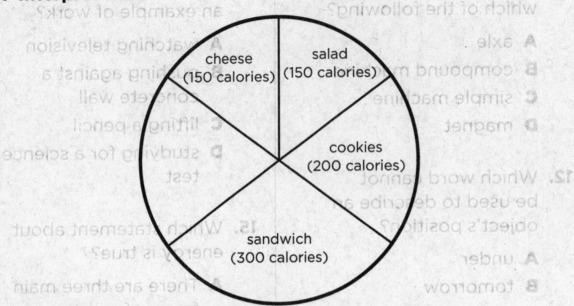

Which of the following foods would complete the chart, making the total 1,000 calories?

A cream cheese (100 calories)   **C** avocado (150 calories)

B roll with butter (200 calories)   **D** banana (85 calories)

17. **Infer** Describe any of the simple machines found on a bicycle.

Answers will vary but may include: wheel and axle; levers

(pedals); pulley (chain).

18. **Communicate** Describe the position of your teacher's desk in relationship to your desk in your classroom.

Answers will vary but should include words such as: in front

of; behind; next to; near, etc.

© Macmillan/McGraw-Hill

Answer the following questions.

19. Would you use oil on a bicycle chain to increase friction or to reduce it? Explain.

Oil will make the surface of the bicycle chain slippery and it
will reduce friction.

20. Look around your classroom. Identify a compound machine and how it is used in the classroom.

Answers will vary. Accept all reasonable responses.

21. **Critical Thinking** Describe how energy moves from one thing to another when a pitcher throws a baseball and a batter hits it.

Possible answer: Energy from the pitcher's body goes into the
baseball when it is thrown. The batter transfers energy from
his body into the bat, and the bat transfers energy to the ball
when it is hit.

22. **Thinking Like a Scientist** What is one way that the potential energy in wood can be changed into another form of energy?

If wood is burned, its stored energy will turn into heat energy.

Machines in Action

Make a Machine Poster

Objective: Students will make a poster that includes a variety of simple and compound machines. Students will identify and label the simple machines that make up the compound machines.

Materials

- poster board or large white construction paper
- markers or crayons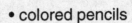
- pencil
- colored pencils
- magazines
- newspapers

Scoring Rubric

4 points Student accurately draws and labels each of the six simple machines. Student shows accurate examples of compound machines and identifies and labels the simple machines that are contained therein. Student clearly explains his or her answers to the questions in Analyze the Results.

3 points Student draws each simple machine, but may have mislabeled one or two machines. Student shows a good representation of compound machines, but may have mislabeled one or two of the simple machines contained therein. Student correctly answers most of the questions in Analyze the Results.

2 points Student draws each simple machine, but may have mislabeled three or four simple machines. Student shows some knowledge of compound machines but mislabels most of the simple machines contained therein. Student attempts to answer the questions in Analyze the Results. Answers to the questions are mostly incorrect.

1 point Student completes part of the activity, but his or her drawings and labels of simple machines are mostly incorrect. Student does not demonstrate a knowledge of compound machines. Student does not answer the questions or most of the answers are incomplete or incorrect.

Machine Poster

Communicate

Use the materials your teacher has given you to make
a poster showing the six simple machines, some
common compound machines, and how we use them
in our lives. Be sure to identify and label all machines.
Also be sure to identify the simple machines that are
contained within any of the compound machines.

Analyze the Results

1. Look at the compound machines you chose for
your poster. What are they best used for?

Answers will vary but may include: scissors, wheelbarrow,

bicycle, and can opener.

2. What are some of the ways simple machines are
used in building houses?

Answers will vary but may include: screws to hold wood

together, pulleys to lift materials, and inclined planes to

move objects.

3. How can using simple machines make heavy
objects seem lighter?

A lever can lift a heavy object on the other side of the lever;

a pulley changes the direction of force to lift an object; an

inclined plane's slanted surface makes it easier to move heavy

objects from one place to another.

Forms of Energy

Write the word or words that best complete each sentence in the spaces below. Words may be used only once.

circuit	insulators	pitch	translucent
conductors	light	temperature	
electric current	opaque	thermal energy	

1. Materials that block light from passing through

   them are _____ opaque _____ .

2. People measure _____ temperature _____ in degrees.

3. Heat moves easily through materials called _____ conductors _____ .

4. Fur, cotton, and wool are examples of _____ insulators _____ .

5. Objects that block some light and let some light

   pass through are _____ translucent _____ .

6. Energy that makes particles move is called _____ thermal energy _____ .

7. A sound's _____ pitch _____ is how high or low it is.

8. A form of energy that allows you to see objects is

   _____ light _____ .

9. A flow of charged particles is called a(n) _____ electric current _____ .

10. A(n) _____ circuit _____ is the path an electric
    current takes.

Circle the letter of the best answer for each question.

11. Which of the following statements is false?

   A Sound is produced by energy of motion.

   B Sound can travel through space.

   C Sound travels fastest through a solid.

   D Sound travels in waves.

12. A black piece of poster board is

   A transparent.

   B translucent.

   C opaque.

   D refractive.

13. What color is the light from the Sun?

   A only red

   B a mixture of many colors

   C only yellow

   D white and red

14. What is static electricity?

   A negatively charged particles

   B a path that an electric current takes

   C a buildup of electrical charge

   D a flow of electrical charge

15. An object that absorbs all light that strikes it would be

   A translucent.

   B transparent.

   C white.

   D black.

Answer the following questions.

16. Interpret Data Study the information in the table below. What color in the visible light spectrum has the longest wavelength?

Electromagnetic Spectrum

Color	Wavelength
violet	400 nanometers
indigo	445 nanometers
blue	475 nanometers
green	510 nanometers
yellow	570 nanometers
orange	590 nanometers
red	650 nanometers

red

17. Infer How does thermal energy affect an object's ability to expand or contract?

When an object gains thermal energy it expands because its

particles move faster and farther apart. When an object loses

thermal energy it contracts because its particles move slower

and closer together.

18. Communicate A clear image cannot be reflected from a rough surface. Why is this true?

Light must hit a smooth surface to reflect uniformly and to

form a clear image. If the surface is rough, then it has lost

some of its ability to reflect light clearly and the light bounces

in many different directions.

© Macmillan/McGraw-Hill

Answer the following questions.

19. Explain why some winter blankets are made of wool.

Heat does not move through wool easily because it is a good

insulator. Wool blankets can trap body heat.

20. How does a light switch control the flow of an electrical current?

When the switch is in the off position, it creates a gap in

the electrical current's path, opening the circuit so that the

current cannot flow. When the switch is in the on position, the

gap is closed and the current can flow through the circuit.

21. Critical Thinking What happens when white light strikes a yellow flower?

All the colors in the white light are absorbed except for

yellow, which is reflected and is the color that can be seen.

22. Thinking Like a Scientist Describe how a thermometer works.

A thermometer is usually made of a glass tube filled with

a liquid, like mercury. When the temperature of the liquid

increases, the particles of the liquid move farther apart.

As the liquid expands, it rises to fill more of the tube. The

temperature is found by comparing the height of the liquid to

the scale provided on the thermometer.

Name _____   Date _____

Circle the letter of the best answer for each question.

1. Which of these is a good conductor?

   A wool

   B cotton

   C metal

   D fur

2. Which of these has the slowest moving particles?

   A water

   B ice

   C a wooden ruler

   D a fire

3. What do you call the flow of energy that moves between objects?

   A temperature

   B heat

   C energy

   D friction

4. What happens when an object gains thermal energy?

   A it cools down

   B it floats

   C it expands

   D it contracts

Critical Thinking If a cool metal spoon is placed into a bowl of hot soup, what would happen first? Would the spoon heat up, or would the soup cool down? Why?

The spoon would heat up before the soup cooled down because

the energy would move from the warmer object to the cooler

object.

© Macmillan/McGraw-Hill

Circle the letter of the best answer for each question.

1. How do all sounds begin?

   A when they reach your ears

   B when something vibrates

   C when they reach a high pitch

   D when they sustain a low pitch

2. Sound travels slowest through which matter?

   A a solid

   B a liquid

   C a gas

   D in space

3. What is vibration?

   A a slow steady sound

   B quick movement back and forth

   C a hard pounding sound

   D how high or low a sound is

4. What is volume?

   A how loud a sound is

   B how high or low a sound is

   C how steady a sound is

   D how much variety a sound has

Critical Thinking How can listening to loud music through headphones possibly damage a person's ability to hear?

Loud sounds have a lot of energy and high energy sounds can
_____
damage the delicate parts inside the ear. Over a period of time,
_____
this can cause hearing loss.
_____

Name _____ Date _____

Circle the letter of the best answer for each question.

1. What is it called when light bounces off an object?

   A absorption

   B refraction

   C vibration

   (D) reflection

2. What does *refract* mean?

   (A) to bend

   B to travel in a straight line

   C to absorb

   D to bounce off

3. Which of these objects absorbs all the light that hits it?

   A a white object

   (B) a black object

   C a mirror

   D a shadow

4. Look at the chart below.

Cause ⟶	Effect
the Sun moves directly overhead at noon ⟶	?

Which of the following belongs in the empty box?

   A your shadow gets longer

   (B) your shadow gets shorter

   C your shadow gets wider

   D your shadow gets darker

Critical Thinking If you wanted to keep sunlight out of a room, what type of shade should you hang—transparent, translucent, or opaque? Why?

I would hang an opaque shade because opaque objects prevent

light from coming through them.

Name _____ Date _____

Circle the letter of the best answer for each question.

1. What is an electrical charge?

  A the positive and negative properties of electricity

  B the uninterrupted flow of electricity

  C a sudden burst of energy

  D the path that allows electrical current to flow

2. What causes a shock when a person touches a doorknob?

  A electrical current

  B a broken circuit

  C static electricity

  D insulated wires

3. People use electricity to produce all of the following except

  A heat.

  B motion.

  C growth.

  D light.

4. Look at the chart below. Which of the following belongs in the empty circle?

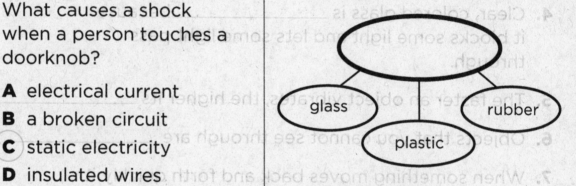

  A Conductors

  B Insulators

  C Reflectors

  D Circuit breakers

Critical Thinking Why are most electrical cords covered in plastic?

Plastic is a good insulator. Without a plastic covering the wires

would be exposed and too hot and dangerous to touch.

_____

© Macmillan/McGraw-Hill

Name _____ Date _____

Forms of Energy

**Write the word that best completes each sentence
in the spaces below. Words may be used only once.**

conductor	pitch	switch	volume
insulator	reflects	translucent	
opaque	shadow	vibrates	

1. A(n) _____ opens and closes a circuit.

2. Heat does not move easily through a(n) _____ .

3. When light bounces off an object it _____ .

4. Clear, colored glass is _____ because
   it blocks some light and lets some light pass
   through.

5. The faster an object vibrates, the higher its _____ .

6. Objects that you cannot see through are _____ .

7. When something moves back and forth quickly it

   _____ .

8. Metal is a good _____ because heat
   moves through it easily.

9. A plane taking off has a greater _____
   than a bird's song.

10. When your body blocks sunlight, you form a(n)

   _____ .

© Macmillan/McGraw-Hill

Answer the following questions.

19. Explain why on a hot day the sand and the water at the seaside are different temperatures.

Even though both are warmed by the Sun, water needs more

energy than sand to cause the same change in temperature.

20. What materials would a manufacturer need to make a wire that is safe to use, but also conducts electricity?

Answers will vary but may include: Copper wire would

conduct electricity; a plastic or rubber coating would insulate

it and make it safe to use.

21. Critical Thinking Explain why clapping your hands is a louder sound than snapping your fingers.

Loud sounds are created when things vibrate with a lot of

energy. Clapping your hands requires more energy than

snapping your fingers and would have a greater volume.

22. Thinking Like a Scientist A scientist needs to use hot test tubes in an experiment. What can she use to hold the hot objects safely? What material should it be made out of?

The scientist needs to use something that conducts heat

poorly and would not allow the heat to transfer to her hand.

She could use a wood or plastic tool to hold the test tubes

safely or a metal tool with an insulated handle.

© Macmillan/McGraw-Hill

Binoculars

Materials

- bathroom tissue rolls
- clear plastic wrap
- rubber bands

Make Binoculars

Objective: Students will construct a set of binoculars to observe opaque, transparent, and translucent objects in the classroom. Students will make a chart describing the objects they viewed with their binoculars. Students can display their charts in the classroom to compare their observations.

Scoring Rubric

4 points   Student correctly assembles a set of binoculars and uses it to record one transparent, translucent, and opaque object in the classroom. Student provides detailed descriptions of these objects in a chart. Student clearly explains his or her answers to the questions in Analyze the Results.

3 points   Student correctly assembles a set of binoculars and uses it to record one transparent, translucent, and opaque object in the classroom. Student successfully makes a chart, but descriptions are limited. Student's answers to questions in Analyze the Results are mostly correct.

2 points   Student assembles a set of binoculars, but it may not be well-designed. Student uses binoculars to record one transparent, translucent, and opaque object in the classroom, but one or more of these objects may be mislabeled. Student records each object on a chart, but does not include descriptions. Student's answers to questions in Analyze the Results are mostly incorrect.

1 point   Student assembles a set of binoculars, but it may not be well-designed or complete. Student records two or fewer objects, and one or more object is mislabeled. Student's chart is incomplete. Student's answers to the questions in Analyze the Results are unclear or incomplete.

Binoculars

Communicate

Use the materials provided by your teacher to make a set of binoculars. Then use the binoculars to look around the classroom to find one opaque object, one transparent object, and one translucent object. Make a chart listing each of the objects you see. On your chart, record details describing your objects, including: color, whether or not it can cast a shadow, what kind of thermal energy it might contain, and any other details you find interesting. Compare your observations with other students.

Analyze the Results

1. Describe how you used the materials given to make a model of binoculars.

   Answers will vary but may include: I placed plastic wrap on

   the ends of the bathroom tissue roll with rubber bands. Then I

   used more rubber bands to keep the rolls together.

2. Why is plastic wrap used? What would happen if wax paper was used instead?

   Plastic wrap is transparent and allows the light to get through

   completely. Wax paper, or other translucent or opaque

   materials, would not allow all of the light to pass through.

3. What color is each object you observed? Why does the object appear as that color?

   A color can be seen because when light hits an object, the

   object will reflect its color and absorb all other colors.

Binoculars

Communicate

Use the materials provided by your teacher to make a set of binoculars. Then use the binoculars to look around the classroom to find one opaque object, one transparent object, and one translucent object. Make a chart listing each of the objects you see. On your chart, record details describing your objects, including color, whether or not it can cast a shadow, what kind of thermal energy it might contain, and any other details you find interesting. Compare your observations with other students.

Analyze the Results

1. Describe how you used the materials given to make a model of binoculars.

_____

_____

_____

2. Why is plastic wrap used? What would happen if wax paper was used instead?

_____

_____

_____

3. What color is each object you observed? Why does the object appear as that color?

_____

_____